Robert DeCourcy Ward

Practical exercises in elementary meteorology

Robert DeCourcy Ward
Practical exercises in elementary meteorology
ISBN/EAN: 9783337276164

Printed in Europe, USA, Canada, Australia, Japan

Cover: Foto ©berggeist007 / pixelio.de

More available books at **www.hansebooks.com**

PRACTICAL EXERCISES

IN

ELEMENTARY METEOROLOGY

BY

ROBERT DeCOURCY WARD
INSTRUCTOR IN CLIMATOLOGY IN HARVARD UNIVERSITY

BOSTON, U.S.A.
GINN & COMPANY, PUBLISHERS
The Athenæum Press
1899

COPYRIGHT, 1899, BY
ROBERT DeCOURCY WARD

ALL RIGHTS RESERVED

PREFACE.

THE advance of meteorology as a school study has been much hampered by the lack of a published outline of work in this subject which may be undertaken during the school years. There are several excellent text-books for more advanced study, but there is no laboratory manual for use in the elementary portions of the science. In many secondary schools some instruction in meteorology is given, and the keeping of meteorological records by the scholars is every year becoming more general. There is yet, however, but little system in this work, and, in consequence, there is little definite result. The object of this book is to supply a guide in the elementary observational and inductive studies in meteorology. This Manual is not intended to replace the text-books, but is designed to prepare the way for their more intelligent use. Simple preliminary exercises in the taking of meteorological observations, and in the study of the daily weather maps, as herein suggested, will lay a good foundation on which later studies, in connection with the text-books, may be built up. Explanations of the various facts discovered through these exercises are not considered to lie within the scope of this book. They may be found in any of the newer text-books.

This Manual lays little claim to originality. Its essential features are based on the recommendations in the Report on Geography of the Committee of Ten. A scheme of laboratory exercises, substantially the same as that proposed in this Report, was, for some fifteen years, the basis of the work in elementary meteorology done in Harvard College under the direction of Professor William M. Davis. The plan proposed by the Committee of Ten has been thoroughly tested by the writer during the past five years, not only in college classes, but also in University Extension work among school teachers, and the present book embodies such modifications of that scheme and additions to it as have been suggested by experience. Emphasis is laid throughout this Manual on the larger lessons to be learned from the individual exercises, and on the relations of various atmospheric phenomena to human life and activities. No attempt is made to specify in exactly what school years this work should be undertaken. At present, and until meteorology attains a recognized position as a school study, teachers must obviously be left to decide this matter according to the opportunities offered in each school. The general outline of the work, however, as herein set forth, is intended to cover the grammar and the high school years, and may readily be adapted by the teacher to fit the circumstances of any particular case.

This book contains specific instructions to the student as to the use of the instruments; the carrying out of meteorological observations; the investigation of special simple problems by means of the instruments; and the practical use of the daily weather maps. The Notes for the Teacher, at the end of the book, are explanatory, and contain suggestions which may be useful in directing the laboratory work of the class.

It has been the privilege of the author during the past ten years to study the science of meteorology, and the methods of teaching that science, under the constant direction of Professor William Morris Davis, of Harvard University. To Professor Davis the author is further indebted for many valuable suggestions in connection with the arrangement and treatment of the subject-matter of this book. Thanks are due also to Mr. William H. Snyder, of Worcester (Mass.) Academy, and to Mr. John W. Smith, Local Forecast Official of the United States Weather Bureau, Boston, Mass., for valued criticisms.

ROBERT DeC. WARD.

HARVARD UNIVERSITY, CAMBRIDGE, MASS.,
    September, 1899.

CONTENTS.

INTRODUCTION.

	PAGE
THE IMPORTANCE OF METEOROLOGY: ITS RELATIONS TO MAN	xi

PART I. — NON-INSTRUMENTAL OBSERVATIONS.

CHAPTER I. — OBSERVATIONS OF TEMPERATURE, WIND DIRECTION AND VELOCITY, STATE OF SKY, AND RAINFALL 1

PART II. — INSTRUMENTAL OBSERVATIONS.

CHAPTER II. — ELEMENTARY INSTRUMENTAL OBSERVATIONS . . 11
CHAPTER III. — ADVANCED INSTRUMENTAL OBSERVATIONS 26

PART III. — EXERCISES IN THE CONSTRUCTION OF WEATHER MAPS.

CHAPTER IV. — THE DAILY WEATHER MAP 47
CHAPTER V. — TEMPERATURE 51
CHAPTER VI. — WINDS 70
CHAPTER VII. — PRESSURE 76
CHAPTER VIII. — WEATHER 85

PART IV. — THE CORRELATIONS OF THE WEATHER ELEMENTS AND WEATHER FORECASTING.

CHAPTER IX. — CORRELATION OF THE DIRECTION OF THE WIND AND THE PRESSURE 91
CHAPTER X. — CORRELATION OF THE VELOCITY OF THE WIND AND THE PRESSURE 93
CHAPTER XI. — FORM AND DIMENSIONS OF CYCLONES AND ANTICYCLONES 96
CHAPTER XII. — CORRELATION OF CYCLONES AND ANTICYCLONES AND THEIR WIND CIRCULATION 98

CONTENTS.

	PAGE
CHAPTER XIII. — CORRELATION OF THE DIRECTION OF THE WIND AND THE TEMPERATURE	101
CHAPTER XIV. — CORRELATION OF CYCLONES AND ANTICYCLONES AND THEIR TEMPERATURES	104
CHAPTER XV. — CORRELATION OF THE DIRECTION OF THE WIND AND THE WEATHER	106
CHAPTER XVI. — CORRELATION OF CYCLONES AND ANTICYCLONES AND THE WEATHER	109
CHAPTER XVII. — PROGRESSION OF CYCLONES AND ANTICYCLONES	111
CHAPTER XVIII. — SEQUENCE OF LOCAL WEATHER CHANGES	113
CHAPTER XIX. — WEATHER FORECASTING	114

PART V. — PROBLEMS IN OBSERVATIONAL METEOROLOGY.

CHAPTER XX. — TEMPERATURE	125
CHAPTER XXI. — WINDS	130
CHAPTER XXII. — HUMIDITY, DEW, AND FROST	132
CHAPTER XXIII. — CLOUDS AND UPPER AIR CURRENTS	136
CHAPTER XXIV. — PRECIPITATION	138
CHAPTER XXV. — PRESSURE	139
CHAPTER XXVI. — METEOROLOGICAL TABLES	142

APPENDIX A.

SUGGESTIONS TO TEACHERS	171

APPENDIX B.

THE EQUIPMENT OF A METEOROLOGICAL LABORATORY	186

INDEX	197

ACKNOWLEDGMENT OF FIGURES.

1, 7, 8, 9, 10, 16. Meteorological Instruments. H. J. Green, 1191 Bedford Avenue, Brooklyn, N. Y.

2, 4. Instrument Shelter and Rain Gauge. *Instructions for Voluntary Observers.* United States Weather Bureau.

5. Mercurial Barometer. L. E. Knott Apparatus Co., 14 Ashburton Place, Boston, Mass.

12, 15, 53. Thermograph and Barograph Curves, and Cyclonic Composite. Davis, *Elementary Meteorology.*

17. Nephoscope. *Annals Harvard College Observatory*, Vol. XX, Part I.

48. North Atlantic Cyclone. *Pilot Chart of the North Atlantic Ocean.* United States Hydrographic Office.

51. Wind Rose. *Quarterly Journal Royal Meteorological Society*, Vol. XXIV, No. 108.

INTRODUCTION.

THE IMPORTANCE OF METEOROLOGY: ITS RELATIONS TO MAN.

WE live in the laboratory of the earth's atmosphere. The changes from hot to cold, wet to dry, clear to cloudy, or the reverse, profoundly affect us. We make and unmake our daily plans; we study or we enjoy vacations; we vary our amusements and our clothing according to these changes. The weather forecasts for the day in the newspaper are read even before the telegraphic despatches of important events. Sailors about to put to sea govern themselves according to the storm warnings of our Weather Bureau. Farmers and shippers of fruit, meat, and vegetables anxiously watch the bulletins of cold or warm waves, and guard against damage by frost or excessive heat. Steam and electric railways prepare their snow-plows when a severe snowstorm is predicted.

Meteorology, the science of the atmosphere, is thus of very great interest and importance. There is no subject a knowledge of which does more to make our daily life interesting. Since we live in the midst of the atmosphere and cannot escape from the changes that take place in it, we must, consciously or unconsciously, become observers of these changes. Examples of the varying processes at work in the atmosphere are always with us. There is no end to the number and the variety of our illustrations of these processes. Man is so profoundly affected by weather changes from day to day that all civilized countries have established weather services. Observers taking regular weather records are stationed at thousands of different places in all parts of the world, and the observations which they make are used by meteorologists in preparing daily weather maps and forecasts, and in studying the conditions of temperature, winds, and rainfall. In the United States alone there are about 3000 of these observers.

These observations are not made on land only. Hundreds of ship captains on all the oceans of the world are making their regular daily meteorological records, which at the end of the voyage are sent to some central office,[1] where they are studied and employed in the preparation of Pilot Charts for the use of mariners. By means of these ocean meteorological observations, which were first systematized and carried out on a large scale under the direction of Lieutenant Matthew Fontaine Maury (born, 1806; died, 1873), of the United States Navy, it has become possible to lay out the most favorable sailing routes for vessels engaged in commerce in all parts of the world.

So important is a knowledge of the conditions of the winds and the weather, that scientific expeditions into unexplored or little-known regions give much of their time to meteorological observations. On the famous Lady Franklin Bay Expedition (1881–1884) of Lieutenant (now General) A. W. Greely, of the United States Army, meteorological observations were kept up by the few feeble survivors, after death by disease and starvation had almost wiped out the party altogether, and when those who were left had but a few hours to live unless rescue came at once. On Nansen's expedition to the "Farthest North," on Peary's trips to Greenland, and on every recent voyage to the Arctic or the Antarctic, meteorological instruments have formed an important part of the equipment.

Not content with obtaining records from the air near the earth's surface, meteorologists have sent up their instruments by means of small, un-manned balloons to heights of 10 miles; and the use of kites for carrying up such instruments

---

[1] In the United States, marine meteorological observations are forwarded to the United States Hydrographic Office, Navy Department, Washington.

has been so successful that, at Blue Hill Observatory, near Boston, Mass., records have been obtained from a height of over 2 miles.* Observatories have also been established on mountain summits, where meteorological observations have been made with more or less regularity. Such observatories are those on Pike's Peak, Colorado (14,134 feet), Mont Blanc, Switzerland (15,780 feet), and on El Misti, in southern Peru. The latter, 19,200 feet above sea level, is the highest meteorological station in the world.

The study of the meteorological conditions prevailing over the earth has thus become of world-wide importance. In the following exercises we shall carry out, in a small way, investigations similar to those which have occupied and are now occupying the attention of meteorologists all over the world.

PRACTICAL EXERCISES IN ELEMENTARY METEOROLOGY.

PART I. — NON-INSTRUMENTAL OBSERVATIONS.

CHAPTER I.

OBSERVATIONS OF TEMPERATURE; WIND DIRECTION AND VELOCITY; STATE OF SKY, AND RAINFALL.

BEFORE beginning observations with the ordinary instruments, accustom yourself to making and recording observations of a general character, such as may be carried out without the use of any instruments whatever. Such records include: *Temperature; Wind Direction and Velocity; State of the Sky, and Rainfall.*

Temperature. — In keeping a record[1] of temperature without the use of a thermometer, excellent practice is given in observations of the temperature actually felt by the human body. Our bodies are not thermometers. They do not indicate, by our sensations of heat or cold, just what is the temperature of the surrounding air, but they try to adjust themselves to the conditions in which they are. This adjustment depends on many things beside the temperature of the air; *e.g.*, the moisture or humidity of the air; the movement of the air; the temperature and the nearness of surrounding objects. In summer, a day on which the temperature reaches 80° or 85° often seems much hotter than another day on which the temperature rises

[1] Each scholar will need a blank book in which to preserve the observations.

to 95°. In winter, temperatures registered by the thermometer as 10° or 15° above zero often feel a great deal colder than temperatures of − 5° or − 10°. In recording your observations on temperature, the record book may be divided into columns as follows : —

SAMPLE RECORD OF TEMPERATURE.

DATE.	HOUR.	TEMPER-ATURE.	REMARKS.
Jan. 16	9 A.M.	Chilly	
" "	12 M.	Warmer	
" "	4 P.M.	"	Growing slowly warmer all day.
" 17.	8 A.M.	Warm	About the same as Jan. 16, 4 P.M.
" "	11 A.M.	Cooler	Began to grow cooler about 10 A.M.
" "	3 P.M.	Colder	Steadily becoming colder.

The following are some of the questions you should ask yourself in carrying out this work. It is not expected that you will be able to answer all these questions at once, but that you will keep them in mind during your studies, and try to discover the answers, as a result of your own observations.

How does it feel to you out of doors to-day? Is it hot, warm, cool, or cold? What is the difference between your feelings yesterday and to-day? Between day before yesterday and to-day? Have you noticed any *regular* change in your feelings as to warmth and cold during three or four successive days? During the past week or two? During the past month? Is there any difference between the temperature of morning, noon, afternoon, and evening? Is there any *regular* variation in temperature during the day? Have there been any *sudden* changes in temperature during the last few days? Have these sudden changes brought warmer or cooler weather? Has the warmer or cooler weather continued for a day or so, or has another change quickly followed the first? Have the sudden changes, if you have noted any, come at any regular times (as morning, afternoon, evening) or at irregular intervals? Does there seem to you to be any definite system, of any kind, in our changes of temperature? In what ways are people in general affected by hot weather? By cold weather? What difference does a very hot or a very cold day make in your own case?

Wind Direction and Velocity. — Wind is an important meteorological element because it has many close relations to human life. It affects very markedly our bodily sensations of heat or cold. A cold, calm day is pleasanter than a cold, windy day. On the other hand, a hot, calm day is usually much more uncomfortable than a hot, windy day. High winds cause wrecks along seacoasts and damage houses, crops, and fruit trees. Sea breezes bring in fresh, cool, pure air from the ocean on hot summer days. In the tropics the sea breeze is so important in preserving the health of Europeans in many places that it is known as " the doctor." The movement of wind through large cities carries off the foul air which has collected in the narrow streets and alleys, and is thus a great purifying agent.

Record the *direction of the wind* according to the four cardinal points of the compass (N., E., S., and W.) and the four intermediate points (NE., SE., SW., and NW.). The direction of the wind is the point *from* which the wind blows. You can determine the points of the compass roughly by noting where the sun rises and where it sets.

Note the *velocity of the wind* according to the following scale, proposed by Professor H. A. Hazen of the United States Weather Bureau.

0 CALM.
1 LIGHT; just moving the leaves of trees.
2 MODERATE; moving branches.
3 BRISK; swaying branches; blowing up dust.
4 HIGH; blowing up twigs from the ground, swaying whole trees.
5 GALE; breaking small branches, loosening bricks on chimneys.
6 HURRICANE or TORNADO; destroying everything in its path.

4   NON-INSTRUMENTAL OBSERVATIONS.

The record book will need two additional columns when wind observations are begun, as follows: —

SAMPLE RECORD OF TEMPERATURE AND WIND.

DATE.	HOUR.	TEMPER-ATURE.	WIND DIRECTION.	WIND VELOCITY.	REMARKS.
Oct. 3	7.30 A.M.	Cool	NE.	Moderate	Temperature falling since last evening. Wind velocity increasing.
" "	11 A.M.	"	"	Brisk	Temperature the same. Wind velocity still increasing.
" "	3 P.M.	"	"	High	Wind velocity still increasing.

What is the direction of the wind to-day? What is its velocity? Has its direction or velocity changed since yesterday? If so, was the change sudden or gradual? Have you noticed any calms? What was the direction of the wind before the calm? What after the calm? Does there seem to be more wind from one compass point than from another? Is there any relation between the direction of the wind and its velocity? *i.e.*, is the NW. wind, for instance, usually a brisk or a high wind, or, is the SE. or S. wind usually moderate? Does the wind usually change its direction gradually, as from SE. to S., then to SW., then to W., etc., or does it jump all at once, as from SE. to W.? Is there any relation between the velocity of the wind and the hour of the day, *i.e.*, does the wind seem stronger or weaker at noon than in the morning or at night? Is it a common occurrence to have a wind from the same direction for several successive days, or are we apt to have different winds almost every day? Do you notice any *systematic* changes in wind direction which are often repeated? What are these changes? Can you make a simple rule for them? In what ways does the wind affect us?

State of the Sky. — By the *state of the sky* is meant the condition of the sky as to its cloudiness. Clouds add much to the beauty and variety of nature. They are often gorgeously colored at sunset. By their changes in form, color, and amount from day to day they relieve what might otherwise be a wearisome succession of the same weather types. Prevailingly overcast skies have a depressing effect. Prevailingly clear skies become monotonous. A proper amount of bright sunshine is essential for the ripening of crops, but too much sunshine may parch soil and vegetation, and become injurious. Clouds bring rain; hence a sufficient amount of cloudiness is just as necessary as a sufficient amount of sunshine. The drift of clouds shows us the direction of movement of the air above us, and is of considerable help in forecasting the weather. Fog, which is a very low cloud, is in some cases so common as to be a meteorological element of great importance. In the city of London, where fogs are very prevalent, especially in winter, the average number of hours of bright sunshine in December and January is only fifteen in each month. The London fogs are, in great part, due to the presence in the air of vast numbers of particles of soot and smoke from millions of fires. These particles increase the density of the fog and prolong its duration.

The amount of cloudiness is recorded on a scale of *tenths*. A *clear* sky is one that is less than $\frac{3}{10}$ cloudy; a *fair* sky is from $\frac{3}{10}$ to $\frac{7}{10}$ cloudy; and a *cloudy* sky is over $\frac{7}{10}$ cloudy. In observing the state of the sky, note such points as the times of clouding and of clearing; the arrangement of the clouds, *i.e.*, whether they are few and scattered, or cover the sky with a uniform layer; the common forms of clouds; the changes in the amounts of cloudiness, etc.

Another new column must be added in the record book for the cloudiness. The table will now appear thus : —

Sample Record of Temperature, Wind, and State of the Sky.

Date.	Hour.	Temperature.	Wind.		State of Sky.	Remarks.
			Direction.	Velocity.		
Dec. 18	9 a.m.	Very cold	NW.	Brisk	Clear	Very cold all night. Everything frozen up.
" "	5 p.m.	" "	"	"	"	Same conditions.
" 19	8.30 a.m.	A little warmer	"	Moderate	Fair	Wind less violent. Small clouds scattered over the sky.

Is the sky *clear*, *fair*, or *cloudy* to-day? Is there more or less cloud than there was yesterday? Than day before yesterday? Is to-day a day of increasing or of decreasing cloudiness? Is the sky usually perfectly clear, or is it oftenest somewhat clouded over? How long does it take for the sky to become completely covered with clouds from the time when it first begins to become cloudy? When there are a few clouds in the sky, are these usually scattered all over the sky, or are they in groups? Have you noticed any particular form of clouds which seemed familiar to you? Do clouds seem to have certain definite shapes and appearances which are to be seen often? Do you discover any variation of cloudiness during the day, *i.e.*, is it apt to be more cloudy in the afternoon than in the morning or at night? Can you make a list describing some of the clouds that you see most often? Can you give these common kinds of clouds some names of your own that shall describe them briefly? In what ways does a clear sky, with bright sunshine, affect us?

Rainfall. — Under the general term *rainfall*, meteorologists include, besides *rain* itself, *snow*, *hail*, *sleet*, etc. The term *precipitation* is also often used. Rainfall stands in close relation to human life and occupations. It feeds lakes and rivers, thus furnishing means of transportation, power for running mills and factories, and water supplies for cities. Regions of abundant rainfall are usually heavily forested, like the Amazon valley in South America, and parts of Equatorial Africa. In civilized countries lumbering is apt to be an important occupation in districts of heavy rainfall, as in Oregon and Washington in our own country, and in Southern Chile in South America. Where there is a moderate rainfall, and other conditions are favorable, there agriculture is possible, and farming becomes one of the chief occupations, as in the Mississippi and Missouri valleys in the United States, and in Western Canada. Districts which have a rainfall too small for successful agriculture, but are not by any means deserts, are often excellent grazing lands, as in the case of parts of Texas, Nebraska, and Kansas in the United States, and the Argentine Republic in South America. Where there is very little rainfall deserts are found. Cities are not built in deserts, because there are no occupations to attract large numbers of men. The inhabitants of the desert are wandering tribes, which move from place to place in search of water and food for themselves and their animals. Rain and snow cleanse the air, washing out impurities such as dust and smoke. Hence they are important agents in preserving health.

Note the *kind* of precipitation (rain, snow, hail, sleet); the *amount* (heavy, moderate, light, trace); and the *time of the beginning and ending* of the storm or shower.

The record book must now be further subdivided into columns, to make room for the rainfall observations, in this manner : —

SAMPLE RECORD OF TEMPERATURE, WIND, STATE OF SKY, AND PRECIPITATION.

DATE.	HOUR.	TEMPER-ATURE.	WIND.		STATE OF SKY.	PRECIPITATION.			REMARKS.
			DIREC-TION.	VELOC-ITY.		TIME OF BE-GINN.	KIND.	AM'T.	
Mar. 21	8.30 A.M.	Mild	S.	Light	Over-cast	8 A.M.	Rain	Light	Raining.
" "	12 M.	"	"	"	Over-cast		"	"	"
" "	4 P.M.	"	"	Moder-ate	Over-cast				Stopped raining about 3 P.M.
" 22	8 A.M.	Cool	NW.	Brisk	Clear				Cleared off during the night.

Does most of our rain come in brief showers, or in storms lasting a day or two? Do we have about the same amount of rain or snow every week and every month, or does the amount vary a good deal from week to week and from month to month? Do you notice much difference in the characteristics of successive storms, or do they all seem pretty much alike? Are thunderstorms limited to any particular season of the year? If so, to what season? Have you discovered any rule as to the time of day when rainstorms or snowstorms begin? When thunderstorms begin and end? Is it common or uncommon for us to have a storm lasting three or four days? How long does a thunderstorm usually last? Do we have most hail in winter or in summer? In what ways does a rainy day affect people? How are you yourself affected? How does a heavy snowstorm affect travel and transportation? In what ways does a snowstorm differ from a rainstorm as to the character of the precipitation and its effects?

After studying the *temperature, wind, state of sky,* and *rainfall* separately, take two elements together and see what relation one has to the other. Try to answer such questions as these: —

Temperature and Wind. — What relations can you discover between the direction of the wind and the temperature? Which winds are the coolest? Which the warmest? Does a hot, calm day seem warmer or cooler than a hot, windy day? Does a cold, calm day seem colder or warmer than a cold, windy day? Does the *velocity* of the wind have any effect on your feeling of cold or of warmth? If so, what effect?

Wind and State of Sky. — Has the direction of the wind anything to do with the cloudiness? Is there more apt to be considerable cloudiness with wind from one direction than from another? What winds are usually accompanied by the largest amount of cloud? What winds usually blow when the sky is clear? Is the relation of cloudiness to certain wind directions so close that, if you know the wind direction, you can make a prediction as to the probable cloudiness? Are the winds with clouds more common in one month than another? In one season than another? If so, which month? which season?

Temperature and State of Sky. — Do you notice any relation between the temperature and the state of the sky? In winter are our coldest days usually cloudy or clear? In summer are our hottest days cloudy or clear? Are the winds that give us the most cloudiness warm or cold winds in winter and in summer? Is a cloudy night colder or warmer than a clear night? Is a cloudy day colder or warmer than a clear day?

State of Sky and Precipitation. — How is rainfall or snowfall related to the cloudiness? Do we ever have rain or snow when the sky is not completely covered with clouds? Does the sky usually become quickly covered with clouds before a rain? Does a sky wholly covered with clouds always give us rain or snow? Does the sky clear rapidly or slowly after a rain? Are any particular kinds of clouds associated with rain or with snowstorms? With brief showers? With thunderstorms?

Wind and Precipitation. — Are any particular wind directions more likely than others to give us rain or snow? Are these the same winds as those which give us the most cloudiness? What winds are they? Has the velocity of the wind any relation to the rain or snowstorm? Does the wind blow harder before, during, or after the rain or snow? What changes of wind direction have you noted before, during, and after any storm? Have you noticed these same changes in other storms? Are they so common in our storms that you can make a rule as to these changes?

Temperature and Precipitation. — Does a shower or a rainstorm in the hotter months affect the temperature of the air in any way? How? In the winter does the temperature show any changes before a snowstorm? Is it usually warmer or colder then than a day or two before the storm and the day after? Is it usually uncomfortably cold during a snowstorm? Are rainy spells in the spring and the autumn months cooler or warmer than clear dry weather?

Part II. — Instrumental Observations.

CHAPTER II.

ELEMENTARY INSTRUMENTAL OBSERVATIONS.

The non-instrumental observations, suggested in the preceding chapter, prepare the way for the more exact records of the weather elements which are obtainable only by the use of instruments. The non-instrumental records are not to be entirely given up, even after the instrumental work and the weather-map exercises have begun, but should be continued throughout the course. Notes on the forms and changes of clouds, on the times of beginning and ending, and on the character of the precipitation, as outlined in the last chapter, and other observations made without the use of instruments, are an essential part of even the most advanced meteorological records.

The simpler instruments are the *ordinary thermometer*, the *wind vane*, the *rain gauge*, and the *mercurial barometer* (in a modified form). Observations with these instruments, although of a simple character, can be made very useful. The advance over the non-instrumental observations, which latter may be termed observations of *sensation*, is a decided one. In place of the vague and untrustworthy statements concerning hot and cold, warm and cool days, we now have actual degrees of temperature to serve as a basis for comparison of day with day or month with month. The measurements of rain and snowfall enable us to study the amounts brought in different storms, the average precipitation of the various months, etc. The important facts of change of pressure now become known, and also the relation of these changes to the weather. Just as we have, in the earlier work, become familiar with our typical weather changes and types, so we shall now have our eyes opened to the actual values of the temperatures and precipitation connected with these changes.

The **ordinary thermometer** (Greek: *heat measure*), the most commonly used and most widely known of all meteorological instruments, was in an elementary form known to Galileo, and was used by him in his lectures at the University of Padua during the years 1592 to 1609. Thermometers enable us to measure the temperatures of different bodies by comparison with certain universally accepted standards of temperature. These standards are the freezing and boiling points of distilled water. In its common form the thermometer consists of a glass tube, closed at the top, and expanding at its lower end into a hollow spherical or cylindrical glass bulb. This bulb and part of the tube are filled with mercury or alcohol. As the temperature rises, the liquid expands, flows out of the bulb, and rises in the tube. As the temperature falls, the mercury or alcohol contracts, and therefore stands at a lower level in the tube. In order that the amount of this rise or fall may be accurately known, some definite scale for measurement must be adopted. The scale commonly used in this country owes its name to Fahrenheit (born in Danzig in 1686; died in 1736), who was the first to settle upon the use of mercury as the liquid in thermometers, and also the first definitely to adopt

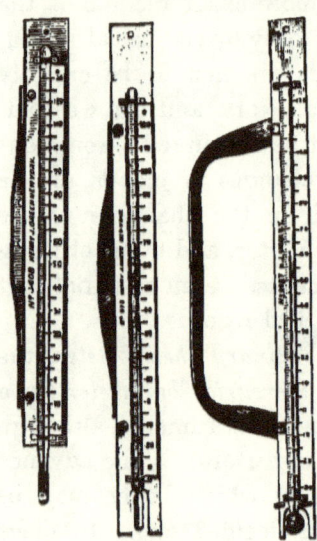

FIG. 1.

two fixed points in graduating the scale. The division of this scale into 180° between the freezing point (32°) and the boiling point (212°) seems to have been taken from the graduation of a semicircle. Fahrenheit was a manufacturer of all sorts of physical apparatus, and it has been thought probable that he had some special facilities for dividing his thermometer tubes into 180 parts. Mercury is most commonly used as the liquid in thermometers, because it readily indicates changes of temperature, and because over most of the world the winter cold is not sufficient to freeze it. The freezing point of mercury is about 40° below zero ($-40°$ F.). Alcohol, which has a much lower freezing point, is therefore used in thermometers which are to be employed in very cold regions. Alcohol thermometers must, for instance, be used in Northern Siberia, where the mean January temperature is 60° below zero.

The temperature which meteorologists desire to obtain by the ordinary thermometer is the *temperature of the free air in the shade.* In order that thermometers may indicate this temperature, they must, if possible, be placed in an open space where there is an unobstructed circulation of the air, and they must be protected from the direct rays of the sun. They are, therefore, usually *exposed* inside of a cubical enclosure of wooden lattice work, in an open space away from buildings, and at a height of 4 to 10 feet above the ground, preferably a grass-covered surface. This enclosure is called the *shelter*, and its object is to screen the instrument from the direct and reflected sunshine, to allow free circulation of air around the bulb, and to keep the thermometer dry. Sometimes the shelter, instead of being in an open space on the ground, is built on the roof or against the north wall of a building, or outside of one of the windows. Fig. 2 shows an ordinary shelter.

A still simpler method of exposure is described in the "Instructions for Voluntary Observers" (United States Weather

Bureau, 1892) as follows: "Select a north window, preferably of an unoccupied room, especially in winter. Fasten the blinds open at right angles to the wall of the house. Fasten a narrow strip 3 inches wide across the window outside, and from 8 to 12 inches from the window-pane. To this fasten the thermometers." If none of these methods of sheltering the instrument is feasible, the thermometer may be fastened to the window frame, about a foot from the window, and so arranged that it can be read from the inside of the room without opening the window.

FIG. 2.

Readings of the thermometer, to the nearest degree of temperature indicated on the scale by the top of the mercury column, are to be made at the regular observation hours, and are to be entered in your record book. Temperatures below zero are preceded by a minus sign (−). A table similar to that suggested towards the close of the last chapter (p. 8) may be used in keeping these instrumental records, except that actual thermometer readings can now be entered in the column headed "Temperature," instead of using only the general terms *warm*, *cold*, *chilly*, etc. This is a great gain. You will now be able to give fairly definite answers to many of the questions asked in Chapter I. Answer these questions with the help of your thermometer readings, as fully as you can.

THE WIND VANE.

The greater part of the Temperate Zone, in which we live, is peculiar in having frequent and rapid changes of temperature, not only from season to season, but from day to day, and during a single day. In winter, we are apt to have a warm wind immediately after a spell of crisp cold weather. In summer, cloudy, cool days come as a sudden relief when we have been suffering from intense heat, with brilliant sunshine.

These changes give a variety to our climate which is, on the whole, very beneficial to man. The North Temperate Zone, with strong seasonal changes in temperature and weather, is the zone of the highest civilization and of the greatest energy of man. In the Torrid Zone the changes of temperature are, as a whole, small. There is no harsh winter. The climate is monotonous and deadening, rather than enlivening. Man finds it easy to live without much work, and the inhabitants of the Torrid Zone have not, as a rule, advanced far in the scale of civilization.

The **wind vane** used by the Weather Bureau is about 6 feet long, and has a divided tail made of pine boards, the two pieces making an angle of 22½°. The purpose of this divided tail is to steady the vane and to make it more sensitive to light currents.

A common wind vane on a neighboring church steeple or flagstaff will usually serve sufficiently well for ordinary use. Observations of wind direction (to eight compass points) are to be made as a part of the ordinary weather record, and to be entered in the proper column of the record book.

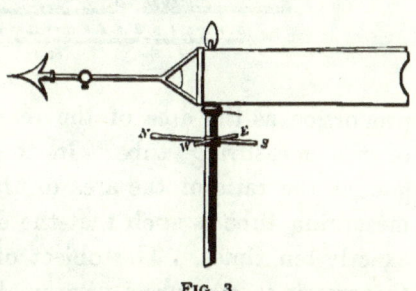

Fig. 3.

The **rain gauge** consists of three separate parts, the receiver A, the overflow attachment B, and the measuring tube C. The inside diameter of the top of the receiver in the standard Weather Bureau gauge is 8 inches (at a in Fig. 4). This receiver has a funnel-shaped bottom, so that all the precipitation which falls into it is carried at once into the measuring tube C, whose inside height is 20 inches. The diameter of the measuring tube is 2.53 inches. The rain falling into the receiver A fills this tube C to a depth greater than the actual rainfall, in

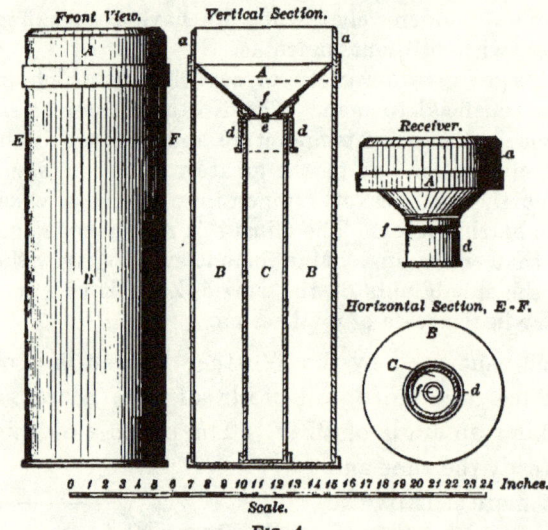

Fig. 4.

proportion as the area of the receiver is greater than the area of the measuring tube. In the standard Weather Bureau gauges the ratio of the area of the receiver to the area of the measuring tube is such that the depth of rainfall is magnified exactly ten times. The object of magnifying the amount in this way is to measure a very small quantity more easily. The narrow portion of the receiver [d] fits over the top of the measuring tube, holding the latter firmly in place and preventing any loss of rainfall. An opening, e, in the lower portion of the receiver [d], just on a level with the top of the measuring tube, serves as an escape for the water into the overflow attachment B, in case the rainfall is so heavy as to more than fill the tube. The inside diameter of the overflow attachment is the same as that of the receiver (8 inches), as will be seen from the figure.

The rain gauge should be firmly set in a wooden frame, so arranged that the overflow attachment can readily be removed from the frame. The box in which the gauge is sent out by the manufacturer is usually designed to serve as a permanent support when the gauge is set up. The best exposure for the gauge is an open space unobstructed by large trees, buildings, or fences. Fences, walls, or trees should be at a distance from the gauge not less than their own height. If an exposure upon the ground is out of the question, the gauge may be placed upon a roof, in which case the middle of a flat unobstructed roof is the best position.

Records of Rainfall. — Every rain gauge is provided with a measuring stick, which is graduated into inches and hundredths. It must be remembered that the amount of rain in the measuring tube is, by the construction of the ordinary gauge, ten times greater than the actual rainfall. This fact need not, however, be taken into account by the observer, for the numbers used in graduating the measuring sticks have all been divided by 10, and therefore they represent the actual rainfall. The graduations on the stick indicate hundredths of an inch, and should appear in the record as decimals (.10, .20, etc.). Ten inches of water in the measuring tube will reach the mark 1.00 on the stick; thus 1.00 denotes 1 inch and zero hundredths of rain. One inch of water in the tube will reach the .10 mark, indicating $\frac{10}{100}$ of an inch. The shortest lines on the measuring stick denote successive hundredths of an inch. Thus, if water collected comes to a point halfway between the .10 and .20 lines, the amount is .15 inch, and so on. In measuring rainfall, the stick is lowered through the bottom of the receiver into the measuring tube, and on being withdrawn the wet portion of the stick at once shows the depth of water in the tube. Care must be exercised to put the end of the stick where the numbering begins first into the gauge, and to pass the stick through the middle of the tube. After each observation the gauge should be emptied and drained, and immediately put back into place. When the total rainfall more than fills the measuring tube, *i.e.*, exceeds 2 inches, the receiver should first be lifted off and the tube removed with great care so as not to spill any water. After emptying the tube, the surplus water in the overflow attachment must be poured into the measuring tube and measured. The amount of rainfall thus found is to be added to the 2 inches contained in the measuring tube in order to give the total rainfall. If any water happens to be spilled during its removal from the overflow attachment, then the amount in the tube will be less than 2 inches, and it must be carefully measured before the latter is emptied.

During the winter season, in all regions where snow forms the chief part of the precipitation, the only portion of the rain gauge that need be exposed is the overflow attachment. The snow which falls into the gauge may be measured by first melting the snow and then measuring the water as rainfall. About 10 inches of snow give, on the average, 1 inch of water, but the ratio varies very greatly according to the density of the snow. Besides the measurement of the melted snow collected in the gauge, it is customary to keep a record of the depth of snowfall in inches, as measured by means of an ordinary foot rule or a yardstick, on some level place where there has been little or no drifting.

Measurements of rain and snowfall are usually made once a day, at 8 p.m., and also at the end of every storm. Enter the amounts of precipitation in the column of the table headed " Amount " and state always whether it is *rain* or *melted snow* that you have measured. When there has been no precipitation since the last observation, an entry of 0.00 should be made in the column of the record book devoted to " Amount of Precipitation." When the amount is too small to measure, the entry T (for *Trace*) should be made.

Continue your non-instrumental record of the time of beginning and ending of the precipitation as before. Whenever it is possible, keep a record of the total amount of precipitation in each storm, noting this under "Remarks." Try to answer such questions as are asked in Chapter I with the help of your instrumental record of the rain and snowfall. Note what depths of snow in different snowstorms are necessary, when melted, to make 1 inch of water.

The Mercurial Barometer. — Air has weight. At sea level this weight amounts to nearly 15 pounds on every square inch of surface. Imagine a layer of water, 34 feet deep, covering the earth. The weight of this water on every square inch of surface would be the same as the weight of the air. Under ordinary circumstances the weight of the air is not noticeable, because air presses equally in all directions, and the pressure within a body is the same as that outside of it. On account of this equal pressure in all directions, we speak of the *pressure* of the air instead of its *weight*. The effects of the air pressure may become apparent when we remove the air from a surface. By working the piston of a pump in a well we may remove the pressure on the surface of the water in the tube of the pump. When this is done, a column of water rises in the tube until the top of this column is about 34 feet above the level of the rest of the water in the well. The pressure of the atmosphere on the water *outside* of the tube holds up this column of water *inside* the tube.

Galileo (1564–1642) first taught that the air has weight. His pupil Torricelli went a step further. Torricelli saw that the column of water, held up by the pressure of the air in the tube of the pump, must exactly balance a similar column of air, reaching from the surface of the water in the well to the top of the atmosphere. The column of water, in other words, exactly replaces this column of air. While working on this subject, Torricelli, in 1643, performed the following experiment. He filled a glass tube, about 3 feet long and closed at one end, with mercury. After filling the tube, he put his finger over the open end and inverted the tube over a vessel containing mercury. When the lower end of the tube was below the surface of the mercury in the dish, he removed his finger. At once the column of mercury fell in the tube until it stood at a height of about 30 inches, leaving a vacant space of 6 inches in the upper part of the tube. This space has since been known as the *Torricellian vacuum*. Torricelli had proved what he had expected, viz., that the height of the column of liquid which replaces and balances an air column of the same size varies with the weight of that liquid. It takes a column of water 34 feet long to balance a similar column of air. It takes a column of mercury only 30 inches long to balance a similar column of air. This, as Torricelli correctly explained, is due to the fact that mercury is so much ($13\frac{1}{2}$ times) heavier than water. The column of water weighs just the same as the column of mercury. Each column exactly balances an air column of similar cross-section. The height of the water or of the mercury is a measure of the weight or pressure of the air. The greater the pressure on the surface of the water in the well, the higher will be the top of the water in the pump. The greater the pressure on the surface of the mercury in the basin, in the experiment of Torricelli, the higher will the mercury column stand in the glass tube. Either water or mercury may be used as the liquid in the barometer. Otto von Guericke (1602–1686), of Magdeburg, constructed a water barometer about 36 feet long, which he attached to the outside wall of his house. This barometer he used for some months, and made predictions of coming weather changes by means of it. A water barometer is, however, a very unwieldy thing to manage, on account of the great length of its tube. Furthermore, water barometers cannot be used in any countries where the temperatures fall to freezing. Mercury is the liquid universally employed in barometers. It is so heavy that only a small column of it is necessary to balance the atmospheric pressure. Therefore a mercurial barometer is portable. Further, mercury does not freeze until the temperature falls to 40° below zero.

Another name which should be mentioned in connection with the barometer is that of Blaise Pascal, who in 1648 fully confirmed Torricelli's results. Pascal saw that if the mercury column is really supported by the weight of the air, the height of that column must be less on the summit of a mountain than at the base, because there is less air over the top of the mountain than at the bottom, and therefore the weight of the air must be less at the summit. To prove this, he asked his brother-in-law Perrier, who lived at Clermont, in France, to carry the Torricellian tube up the Puy-de-Dôme, a mountain somewhat over 3500 feet high in Central France. This Perrier did on Sept. 19, 1648, and he found, as predicted by Pascal, that the mercury fell steadily in the tube as he went up the mountain, and that at the top of the mountain the column of mercury was over 3 inches shorter than at the base.

The pressure of the atmosphere is a weather element which, unlike the other elements already considered, cannot be observed without an instrument. We cannot, under ordinary conditions at sea level, determine by any of our senses whether the pressure is rising or falling, or is stationary. The pressure on the upper floors of one of our high buildings is shown by a barometer to be considerably lower than it is at the level of the street below, and yet we notice no difference in our feelings at the two levels.

It is only when we ascend far into the air, as in climbing a high mountain or in a balloon, that the much-diminished pressure at these great heights perceptibly influences the human body. Mountain climbers and aëronauts who reach altitudes of 15,000 to 20,000 feet or more, usually suffer from headache, nausea, and faintness, which have their cause in the reduced pressure encountered at these heights.

22   INSTRUMENTAL OBSERVATIONS.

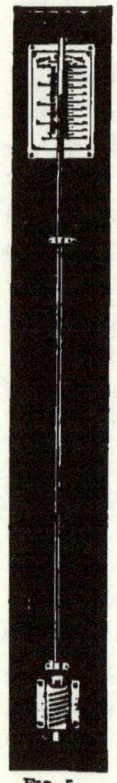

Fig. 5.

The ordinary mercurial barometer in use to-day is, essentially, nothing more than the glass tube and vessel of Torricelli's famous experiment. A simple form of the mercurial barometer is shown in Fig. 5. It consists of a glass tube about one-quarter of an inch in inside diameter and about 36 inches long. This tube, closed at the top and open at the bottom, is filled with mercury, the lower, open end dipping into a cup of mercury known as the *cistern*. The space above the mercury is a vacuum. The mercury extends inside the tube to a height corresponding to the weight or pressure of the air, the vertical height of the top of the mercury column above the level of the mercury in the cistern, in inches and hundredths of an inch, being the barometer reading. At sea level the normal barometer reading is about 30 inches. There is an opening near the top of the cistern, at the back of the instrument, through which the air gains access to the mercury and holds up the mercury column. It will readily be seen that, as the mercury in the tube rises, the level of the mercury in the cistern falls, and *vice versa*, so that there is a varying relation between the two levels. In order to have the reading accurate, it is necessary that the surface of the mercury in the cistern should be just at the zero of the barometer scale when a reading is made. To accomplish this, the bottom of the cistern consists of a buckskin bag which may be raised or lowered by means of a thumb-screw, seen at the lower end of the instrument. The level of the mercury may thus be changed and adjusted to the top of a black line, marked on the outside of the cistern, and which indicates the zero of the scale. Before making a reading, the surface of the mercury in the cistern must be raised or lowered until it *just* reaches this black line. Then the top of the mercury column will give the pressure of the air. The reading is made on an aluminum scale at the top of the wooden back on which the tube is mounted, this scale being graduated both on the English and on the metric system. This barometer may be hung against the wall of a room.

The **aneroid barometer** (Greek: *without fluid*), although less desirable in many ways than the mercurial, is nevertheless a useful instrument for rough observations. The aneroid is not good for careful scientific work, because its readings are apt to be rather inaccurate. To be of much value in indicating exact pressures, it should frequently be compared with and adjusted to a mercurial barometer. An ordinary aneroid barometer is shown in Fig. 6.

Fig. 6.

In this instrument the changes in atmospheric pressure are measured by their effects in altering the shape of a small metallic box, known as the *vacuum chamber*. The upper and lower surfaces of this box are made of thin circular sheets of corrugated German silver, soldered together around their outer edges, thus forming a short cylinder. From this the air is exhausted, and it is then hermetically sealed. A strong steel spring, inside or outside of the vacuum chamber, holds apart the corrugated surfaces, which tend to collapse, owing to the pressure of the external air upon them. An increase or decrease in the air pressure is accompanied by an approach, or a drawing apart, of the surfaces of the chamber. These slight movements are magnified by means of levers, a chain, and a spindle, and are made to turn an index hand or pointer on the face of the instrument. The outer margin of the face, underneath the glass, is graduated into inches and hundredths, and the pressure may thus be read at once.

As the tension of the steel spring varies with the temperature, aneroids are usually *compensated* for temperature by having one of the levers made of two different metals, *e.g.*, brass and iron, soldered together, or else by leaving a small quantity of air in the vacuum chamber. This air, when heated, expands, and thus tends to compensate for the weaker action of the spring, due to the higher temperature. At best, however, this compensation is but imperfect, and this fact, together with the friction of the different parts, the changes in the spring with age, and the need of frequent adjustments, makes aneroids rather inaccurate. They may be adjusted to mercurial barometers by means of a small screw, whose head may be found on the lower surface of the instrument. The words *fair*, *stormy*, etc., which frequently appear on the face of aneroid barometers, are of little use in foretelling weather changes, as no definite pressures always occur with the same weather conditions. The instrument should be tapped lightly a few times with the finger before a reading is made. The second pointer, which is often found in aneroids, is set by the observer on the position marked by the index hand when he makes his reading. The difference between the pressure marked by this set pointer and that shown by the index hand at the next observation is the measure of the change of pressure in the interval.

Another column must now be added to the record book (preferably between the columns devoted to *temperature* and *wind*) to receive the " Pressure in Inches and Hundredths."

Is the pressure constant (*i.e.*, are the readings always the same) or does it vary? If it varies, is there any apparent system in the variations? Is there a tendency to a daily maximum? To a daily minimum? If so, about what time do these occur, respectively? What is the average variation (in inches and hundredths) in the course of a day? What is the greatest difference in pressure which you have observed in a day? What is the least? Does the pressure seem to vary more or less in the colder months than in the warmer? Has the height of the mercury column any relation to the weather? Are we likely to have rainy weather with rising barometer? Is the velocity of the wind related to the pressure in any way? How? Can you make any general rules for weather prediction based on the action of the barometer? What rules?

Tabulation of Observations. — The tables suggested in the preceding chapter can be used unchanged with the simple instruments just described.

Summary of Observations. — At the end of each month summarize your instrumental observations in the following way : —

Temperature. — Add together all your temperature readings ; divide their sum by the total number of observations of temperature, and the quotient will give you a sufficiently accurate *mean* or *average* temperature for the month in question. It is to be noted that the mean monthly temperatures obtained from these observations will be much more accurate if the thermometer readings are made at 7 A.M. and 7 P.M., at 8 A.M. and 8 P.M., etc., and the mean of these is taken ; or if the mean is derived from the maximum and the minimum temperatures, discussed in Chapter III. This *mean* temperature should be written at the bottom of the temperature column, and marked " Mean." The mean monthly temperature is one of the important meteorological data in considering the climatic conditions of any place.

Wind. — Determine the frequency of the different wind directions by counting the total number of times the wind has blown from N., NE., E., etc., during the month. The wind which you have observed the greatest number of times is the *prevailing* wind. It may, of course, happen that two or three directions have been observed an equal number of times. The number of calms should also be recorded.

Rainfall. — The total monthly precipitation is obtained by adding together all the separate amounts of rainfall noted in your record book, and expressing the total, in inches and hundredths, at the bottom of the rainfall column. You now have the means for comparing one month's rainfall with that of another month, and of seeing how these amounts vary.

Examine carefully also your *non-instrumental observations.* See whether you can draw any general conclusions as to the greater prevalence of cloud, or of rain or snow, in one month than in another. Did the last month have more high winds than the one before? Or than the average? Were the temperature changes more sudden and marked? Was there more or less precipitation than in previous months?

CHAPTER III.

ADVANCED INSTRUMENTAL OBSERVATIONS.

THE instruments for more advanced study are the following: *maximum and minimum thermometers, wet and dry bulb thermometers, sling psychrometer, standard barometer, thermograph, barograph, and anemometer.*

Maximum and minimum thermometers are usually mounted together on a board, as shown in Fig. 7, the lower one of the

FIG. 7.

two being the maximum, and the upper the minimum. In the view of the instrument shelter (Fig. 2), these thermometers are seen on the left. The minimum thermometer, when attached to its support, is either exactly horizontal or else slopes downward somewhat towards the bulb end, as shown in Fig. 7. These instruments, as their names imply, register the highest and the lowest temperatures, respectively, which occur during each day of 24 hours. The maximum thermometer is filled with mercury. Its tube is narrowed just above the bulb, in such a way that the mercury passes through the constriction with some difficulty. As the temperature rises, the mercury, in expanding, is forced out from the bulb through this narrow passage. When the temperature falls, however, the mercury above this point cannot get back into the bulb, there being nothing to force it back. The length of the mercury column, therefore, remains the same as it was when the temperature was highest, and the instrument is read by observing the number of degrees indicated by the top, or right-hand end, of the mercury column upon the scale. After reading, the thermometer is set by removing the brass pin upon which the bulb end rests, and whirling the instrument rapidly around the pin to which its upper end is fastened. By this process the mercury is driven back into the bulb, past the constriction. Care must be taken to stop the thermometer safely while it is whirling. After setting, the reading of the maximum thermometer should agree closely with that of the ordinary or dry-bulb thermometer.

The *minimum thermometer* is filled with alcohol, and contains within its tube a small black object, called the *index*, which resembles a double-headed black pin. The instrument is so constructed that this index, when placed with its upper, or right-hand end, at the surface of the alcohol, is left behind within the alcohol, when the temperature rises. On the other hand, when the temperature falls, the index is drawn towards the bulb by the surface cohesion of the alcohol, the top or right end of the index thus marking the lowest temperature reached. The upper end of the thermometer is firmly fastened, by means of a screw, to a brass support, while the lower end rests upon a notched arm. In setting this instrument, the bulb end is raised until the index slides along the tube to the end of the alcohol column. The thermometer is then carefully lowered back into the notch just referred to. Maximum and minimum thermometers need to be read only once a day, in the evening. The temperatures then recorded are the highest and lowest reached during the preceding 24 hours. The observation hour is preferably 8 P.M., but if this is inconvenient, or impracticable, the reading may be made earlier in the afternoon. The hour, however, should be as late as possible, and should not be varied from day to day. The maximum temperature sometimes occurs in the night. The maximum and the minimum temperatures should be entered every day, in a column headed "Maximum and Minimum Temperatures," in your record book.

The **wet and dry bulb thermometers**, together commonly known as the *psychrometer* (Greek: *cold measure*), are simply two ordinary mercurial thermometers, the bulb of one of which is wrapped in muslin, and kept moist by means of a wick leading from the muslin cover to a small vessel of water attached to the frame (see Fig. 8). The wick carries water to the bulb just as a lamp wick carries oil to the flame. The psychrometer is seen inside the shelter on the right in Fig. 2.

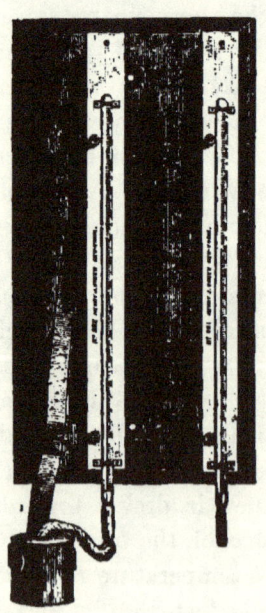

Fig. 8.

The air always has more or less moisture in it. Even the hot, dry air of deserts contains some moisture. This moisture is either invisible or visible. When invisible it is known as *water vapor*, and is a gas. When visible, it appears as *clouds* and *fog*, or in the liquid or solid form of *rain*, *snow*, and *hail*. The amount of moisture in the air, or the *humidity* of the air, varies according to the temperature and other conditions. When the air contains as much water vapor as it can hold, it is said to be *saturated*. Its humidity is then high. When the air is not saturated, evaporation goes on into it from moist surfaces and from plants. Water which changes to vapor is said to *evaporate*.

This process of evaporation needs energy to carry it on, and this energy often comes from the heat of some neighboring body. When you fan yourself on a very hot day in summer, the evaporation of the moisture on your face takes away some of the heat from the skin, and you feel cooler. The drier the air on a hot day, the greater is the evaporation from all moist bodies, and hence the greater the amount of cooling of the surfaces of those bodies. For this reason a hot day in summer, when the air is comparatively dry, that is, not saturated with moisture, is cooler, other things being equal, than a hot day when the air is very moist. Over deserts the air is often so hot and dry that evaporation from the face and hands is very great, and the skin is burned and blistered. Over the oceans, near the equator, the air is hot and excessively damp, so that there is hardly any cooling of the body by evaporation, and the conditions are very uncomfortable. This region is known as the "Doldrums."

The temperatures that are felt at the surface of the skin, especially where the skin is exposed, as on the face and hands, have been named *sensible temperatures*. Our sense of comfort in hot weather depends on the *sensible* temperatures. These sensible temperatures are not the same as the readings of the ordinary (dry-bulb) thermometer, because our sensation of heat or cold depends very largely on the amount of evaporation from the surface of the body, and the temperature of evaporation is obtained by means of the wet-bulb thermometer. Wet-bulb readings at the various stations of the Weather Bureau are entered on all our daily weather maps. In summer (July) the sensible (wet-bulb) temperatures are 20° below the ordinary air temperature in the dry southwestern portion of the United States (Nevada, Arizona, Utah). The mean July sensible temperatures there are from 50° to 65°; while on the Atlantic coast, from Boston to South Carolina, they are between 65° and 75°. Hence over the latter district the temperatures actually experienced in July average higher than in the former.

Unless the air is saturated with water vapor, the evaporation from the surface of the wet-bulb thermometer will lower the temperature indicated by that instrument below that shown by the dry-bulb thermometer next to it, from which there is no evaporation. The drier the air, the greater the evaporation, and therefore the greater the difference between the readings of the two thermometers. By means of tables, constructed on the basis of laboratory experiments, we may, knowing the readings of the wet and dry bulb thermometers, easily determine the *dew-point* and the *relative humidity* of the air — important factors in meteorological observations (see Chapter XXVI). In winter, when the temperature is below freezing, the muslin of the wet-bulb thermometer should be moistened with water a little while before a reading is to be made. The amount of water vapor which air can contain depends on the temperature of the air. The higher the temperature, the greater is the capacity of the air for water vapor. Hence it follows that, if the temperature is lowered when air is saturated, the capacity of the air is diminished. This means that the air can no longer contain the same amount of moisture (invisible water vapor) as before. Part of this moisture is therefore changed, *condensed*, as it is said, from the condition of water vapor into that of cloud, fog, rain, or snow. The temperature at which this change begins is called the *dew-point* of the air.

The *relative humidity* of the air is the ratio between the amount of water vapor which the air contains at any particular time and the total amount which it could contain at the temperature it then has. Relative humidity is expressed in percentages. Thus, air with a relative humidity of 50% has just half as much water vapor in it as it *could* hold.

It is found that the readings of the wet-bulb thermometer are considerably affected by the amount of air movement past the bulb, and that in a light breeze, or in a calm, the reading does not give accurate results as to the humidity of the general body of air outside the shelter.

To overcome this difficulty another form of psychrometer has been devised.

The **sling psychrometer** (Fig. 9) consists simply of a pair of wet and dry bulb thermometers, fastened together on a board or a strip of metal, to the upper part of which a cord with a loop at the end is attached. In this form of psychrometer there is no vessel of water and no wick, but the muslin cover of the wet-bulb thermometer must be thoroughly wet, by immersion in water, just before each observation. The instrument is then whirled around the hand at the rate of about 12 feet a second. After whirling about 50 times, note the readings, and then whirl the instrument again, and so on, until the wet bulb reaches its lowest reading. The lowest reading of the wet bulb, and the reading of the dry bulb at the same time, are the two observations that should be recorded. Take care to have the muslin wet throughout each observation, and' in windy weather stand to leeward of the instrument, so that it may not be affected by the heat of your body. The true reading may be obtained within two or three minutes.

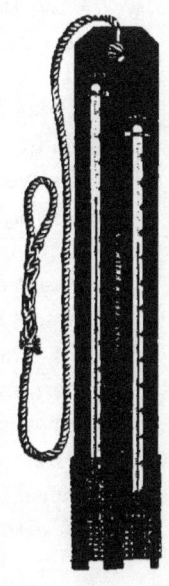

FIG. 9.

Make observations with the wet-bulb thermometer or the sling psychrometer as a part of your regular daily weather record. Note the temperatures indicated by the wet and dry bulbs, and, by means of the table in Chapter XXVI, obtain the *dew-point* and the *relative humidity* of the air at each observation. Enter these data in your record book, in a column headed "Humidity," and subdivided into two columns, one for the dew-point and one for the relative humidity.

By means of observations with the psychrometer you will be able to answer such questions as the following: —

Does the relative humidity vary from day to day? Has it any relation to the direction of the wind? To the state of the sky? To precipitation? Does it show any *regular* variations during the course of a day? How does a high degree of relative humidity affect you in cold weather? In hot weather? Between what limits of percentages does the relative humidity vary? Do the changes come gradually or suddenly? Are these changes related in any way to the changes in the other weather elements? How do the sensible temperatures vary? In what weather conditions do the sensible temperatures differ most from the air temperatures? In what seasons? Compare the sensible temperatures obtained by your own observations with the sensible temperatures at various stations of the Weather Bureau, as given on the daily weather map. Are there any fairly regular differences between the sensible temperatures observed at your own station and the Weather Bureau stations?

Standard Mercurial Barometer. — A simple form of barometer has been described in Chapter II. The ordinary standard mercurial barometer used by the Weather Bureau (Fig. 10) has the glass tube containing the mercury surrounded by a thin brass covering, through which openings are cut, near the top, on the front and back, exposing to view the glass tube and the top of the mercury column. On one side of this opening there is a strip of metal, graduated to inches and tenths or twentieths, by means of which the height of the barometer is determined. This strip, for barometers used at or near sea level, is about 4 inches long, the variations in pressure under normal conditions not exceeding that amount. In addition to this fixed scale there is a small scale, also graduated, which may be moved up and down the opening in the enclosing brass case by means of a milled head outside and a small rack and pinion inside the brass case. This movable scale, known as the *vernier* from the name of its inventor, Vernier, is an ingenious device, by means of which more accurate readings of the barometer can be made than is possible with the ordinary fixed scale. A *vernier* graduated into twenty-five parts enables the observer to make readings accurately to the one-thousandth part of an inch. On the front of the barometer there is a small thermometer, known as the *attached thermometer*. The bulb of this thermometer, concealed within the metal casing of the barometer, is nearly in contact with the glass tube containing the mercury. The air, upon whose weight the height of the mercury column depends, gains access to the top of the cistern through leather joints, by which the cistern is joined to the glass tube.

Mercurial barometers of the Weather Bureau pattern are best hung in a barometer box, fastened securely against the wall of a room, where there is a good light on the instrument and where the temperature is as constant as possible.

In all accurate work certain corrections have to be applied to barometer readings to make them strictly comparable. These are: (1) *correction for altitude;* (2) *correction for temperature;* and (3) *correction for latitude.* The first is necessary because of the fact that the weight of the air decreases upwards, and a barometer reading on a hill or a mountain is not comparable with one at sea level unless the former has been corrected by the addition of the weight of the column of air between the hill or mountain and sea level. The correction for temperature is rendered necessary by the fact that with increasing temperature the mercury in the barometer tube expands more than the metallic scale, because mercury is more sensitive to heat, and unless some allowance is made for this fact, barometer readings made at high temperatures will show somewhat too high a pressure. The readings of the attached thermometer give the temperature of the mercury and are used in making the corrections for temperature. As gravity varies from a maximum value at the poles to a minimum value at the equator, barometer readings made at different latitudes are *corrected for latitude*, which means that they are reduced to latitude 45°, midway between 0° and 90°. The correction is $+0.08''$ at the poles and $-0.08''$ at the equator. Tables for use in correcting barometer readings for altitude and for temperature are given in Chapter XXVI.

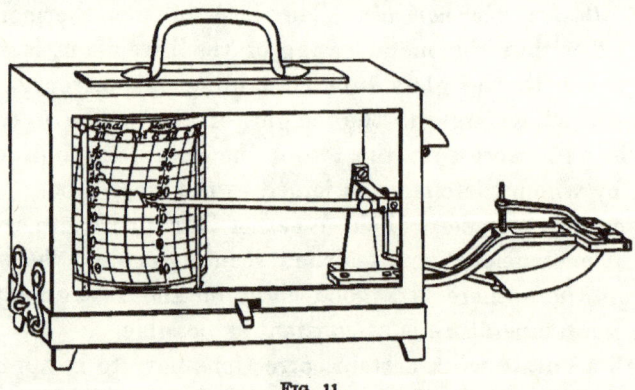

FIG. 11.

Thermograph and Barograph. — Two instruments of much interest are the self-recording thermometer, or *thermograph*, and the self-recording barometer, or *barograph*, manufactured by

Richard Brothers of Paris. In the thermograph (Fig. 11) there is a brass cylinder around which a sheet of paper is wound, this paper being divided into two-hour intervals of time and into spaces representing differences of 5° or 10° of temperature. The cylinder revolves once in a week, being driven by clock-work contained within it. The thermometer consists of a flat, bent, hollow brass tube containing alcohol, one end of the tube being fastened to the metallic frame seen at the right of the figure, and the other end being free to move. With rising temperature, the liquid in the tube expanding more than the metallic casing, by reason of its greater sensitiveness to heat, tends to straighten the tube, while with falling temperature the elasticity of the tube turns it into a sharper curve. These movements of the free end of the tube are carried through a train of levers and thus magnified. At the end of the last lever is a metallic pen filled with ink, which rests lightly against the paper on the revolving drum. A rise or fall in temperature is thus recorded by a rise or fall of the pen on the record sheet, and a continuous curve of temperature is secured. The pen of the thermograph should be frequently adjusted to make the reading of the instrument accord with that of a standard mercurial thermometer, and care should be taken to have the clock keep good time. These adjustments can readily be made by means of a screw and a regulator, respectively. The thermograph should be exposed in the instrument shelter with the other thermometers. The sheets should be changed, the clock wound, and the pen filled once a week, preferably on Monday, at 8 A.M., or at noon.

The continuous records written by a thermograph are a valuable addition to the fragmentary observations which are the result of eye readings of the ordinary thermometer. From the former any omitted thermometer readings may be supplied. The interest of thermograph records may be seen in the following figure (Fig. 12), in which curves traced under different conditions are reproduced. Curve *a* illustrates a period of clear warming weather at Nashua, N. H., April 27–30, 1889. Curve *b* was traced during a spell of cloudy weather at Nashua,

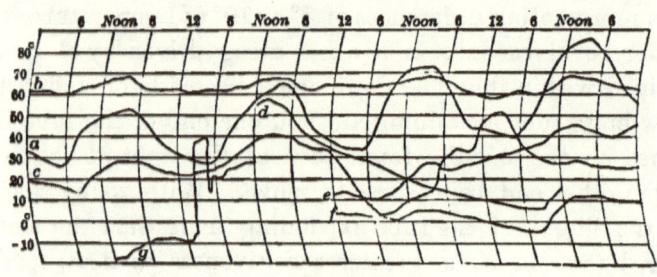

FIG. 12.

accompanying the passage of a West India hurricane, Sept. 13–16, 1889. Curve *c* illustrates the change from a time of moderate winter weather to a cold spell (Nashua, Feb. 22–25, 1889). Curve *d* exhibits a steady fall of temperature from the night of one day over the next noon to the following night, during the approach of a winter cold spell (Nashua, Jan. 19–21, 1889). Curve *e* shows a reverse condition, viz., a continuous rise of temperature through a night from noon to noon (Nashua, Dec. 16–17, 1888). Curve *f* shows the occurrence of a high temperature at night, caused by warm southerly winds, followed by cold westerly winds (Cambridge, Mass., Nov. 30–Dec. 1, 1890). Curve *g* illustrates the sudden rise of temperature due to the coming of a hot, dry wind (chinook) at Fort Assiniboine, Mont. (Jan. 19, 1892). A study of such records leads to the discovery of many important facts, which would be completely lost sight of without a continuous record.

The **barograph** (Fig. 13) is very similar to the thermograph in general appearance. The essential portion of this instrument consists of a series of six or eight hollow shells of corrugated metal screwed one over the other in a vertical column. These shells are exhausted of air, and form, in reality, an aneroid barometer which is six or eight times as sensitive as the ordinary single-chamber aneroid. The springs for distending the shells are inside. The base of the column being fixed, the upper end rises and falls with the variations in

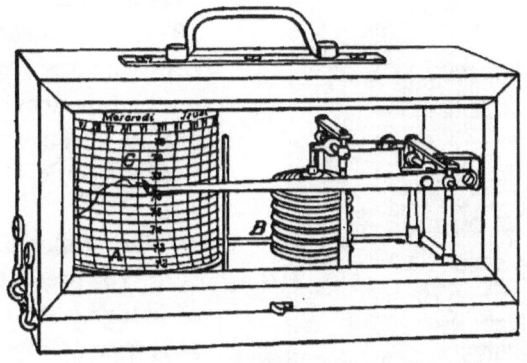

FIG. 13.

pressure. The movements of the shells are magnified by being carried through a series of levers, and, as in the thermograph, the motion is finally given to a pen at the end of the long lever. The compensation for temperature is the same as in the ordinary aneroid. A small quantity of air is left in one of the shells to counteract, by its own expansion at increased temperature, the tendency of the barometer to register too low on account of the weakening of the springs. The barograph may be placed upon a shelf in the school-room, where it can remain free from disturbance, and yet where the record may be clearly seen. The general care of the barograph is the same as that of the thermograph. Brief instructions concerning the care and adjustments of these instruments are sent out by the makers with each instrument. Frequent comparison with a mercurial barometer is necessary, the adjustment of the barograph being made by turning a screw, underneath the column of shells, on the lower side of the wooden case.

Barograph records are fully as interesting as those made by the thermograph. The week's record traced on the writer's

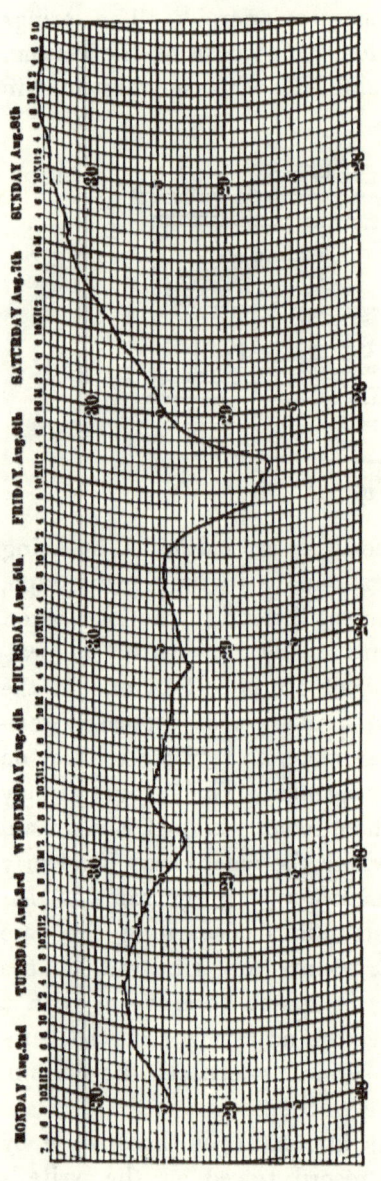

Fig. 14

barograph during a winter voyage from Punta Arenas, Strait of Magellan, to Corral, Chile, Aug. 2–9, 1897, gives a striking picture of the rapid and marked changes of pressure during seven days in the South Pacific Ocean (Fig. 14).

The following figure (Fig. 15) presents samples of barograph curves traced at Harvard College Observatory, Cambridge, Mass., during Feb. 22–28, 1887, and May 17–23, 1887. The February curve illustrates well the large and irregular fluctuations in pressure, characteristic of our winter months; while the May curve shows clearly the more even quality of the pressure changes in our summer.

The **anemometer** shown in Fig. 16 is the most generally used of instruments designed to measure wind velocity. It is known as the Robinson cup anemometer, and consists of four hollow hemispherical cups upon arms crossed at right angles, and all facing the same way around the circle. The cross-arms are fixed

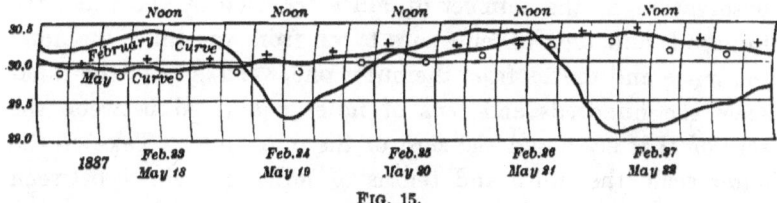

FIG. 15.

upon a vertical axis having an endless screw at its lower end. When the cups move around, the endless screw turns two dials which register the number of miles traveled by

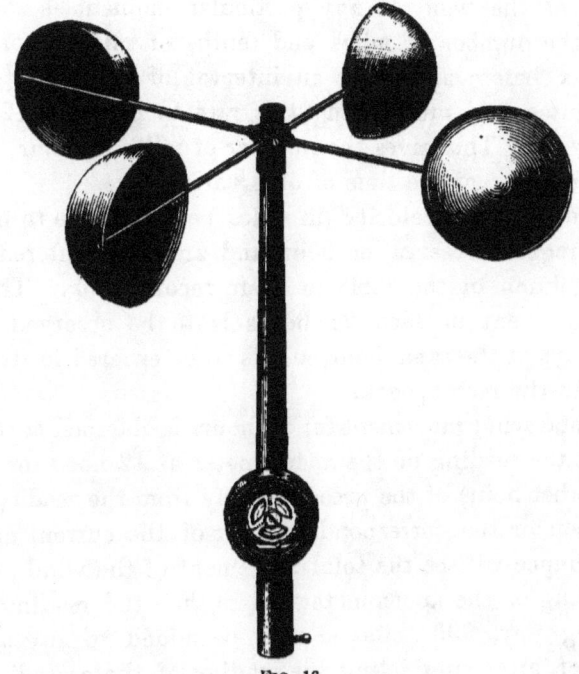

FIG. 16.

the wind. The Weather Bureau pattern of anemometer has the dials mounted concentrically, the outer dial having

100, and the inner, 99 divisions. The revolutions of the outer dial are recorded on the inner one, and in making an observation of the number of miles traveled by the wind, the hundreds and tens of miles are taken from the inner dial, and the miles and tenths from the outer one. Take from the inner scale the hundreds and tens of miles contained between the zero of that scale and the zero of the outer one. Take on the outer scale the miles and tenths of miles contained between the zero of that scale and the index point of the instrument. The sum of these readings is the reading of the instrument at the time of the observation.

Wind velocities are recorded in miles per hour. The velocity of the wind at any particular moment is found by noting the number of miles and tenths of miles recorded by the index before and after an interval of one minute, or of five minutes, and multiplying this rate by 60 or by 12 as the case may be. This gives the number of miles an hour that the wind is blowing at the time of observation.

Records of wind velocity (in miles per hour) are to be made at each regular observation hour, and are to be entered in the proper column of the table in your record book. The total wind movement in each 24 hours is to be observed once a day, always at the same hour, and is to be entered in its proper column in the record book.

The total wind movement for 24 hours is obtained as follows: Subtract the reading of the anemometer at 12 noon (or 8 A.M., or any other hour) of the preceding day from the reading taken at 12 noon or the corresponding hour of the current day, and the difference will be the total movement of the wind. When the reading of the anemometer is less than the reading of the preceding day, 990 miles should be added to it; and the remainder, after subtracting the reading of the preceding day, will be the total wind movement for the 24 hours. Thus: To-day's reading = 91 miles; yesterday's reading = 950 miles.

NEPHOSCOPE.

Hence $91 + 990 = 1081$ miles, $1081 - 950 = 131$ miles = total wind movement for the current day.

By means of an electrical attachment the anemometer may be arranged so as to record continuously on a cylinder rotating by clock-work, a pen making a mark on the paper for every mile traveled by the wind. The anemometer should be exposed on top of a building where there is as little obstruction as possible by tall chimneys, higher buildings, and the like.

The **nephoscope** (Greek: *cloud observer*) is an instrument used in determining the directions of movement of clouds. These directions, if determined by ordinary eye observation of the clouds as they drift across the sky, are apt to be quite inaccurate. The best method of observing directions of cloud movement is to note the path of the reflection of the cloud in a horizontal mirror, the observer looking at this reflection through an eyepiece which remains fixed during the operation. Such a horizontal mirror, adapted to measure the direc-

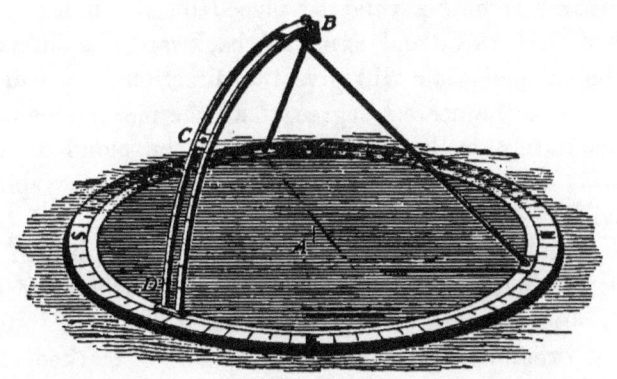

FIG. 17.

tion of motion of clouds, is known as a *nephoscope*. A form of nephoscope devised by Mr. H. H. Clayton, of Blue Hill Observatory, Hyde Park, Mass., is shown in Fig. 17.

This instrument consists of a circular mirror, 13 inches in

diameter, sunk in a narrow circular wooden frame, on top of which is fastened a brass circle, S.W.N.E., divided to 5° of arc. Inside of this fixed circle is a movable brass one, to which is attached a brass arc, BD, rising above the mirror and bearing a movable eyepiece, C. This arc forms the quadrant of a circle whose center is the center of the mirror, and is divided to 5° of arc. Its top is held vertically over the center of the mirror by two rods fastened to the movable circle. The center of the mirror A is marked by cross lines on the reflecting surface, the glass of which is thin. In order to determine the motion of a cloud, the movable circle and tripod are revolved until the arc BD is in the vertical plane formed by the cloud, the center of the mirror, and the eye. The eyepiece C is then shifted until some point of the cloud image, as seen through the eyepiece, is projected on the intersection of the cross lines on the glass. The cloud image soon changes its position, and while the eye is still held at the eyepiece, a small index is placed on the part of the cloud image which previously appeared on the center of the mirror. If now a ruler be placed on the index and the center of the mirror and extended backward, its intersection with the divided scale will give the direction from which the cloud came to the nearest degree, if all the measurements have been accurately made. The height of the cloud above the horizon is found by reading the position of the eyepiece on the divided quadrant.

The nephoscope may be placed on a table, out of doors in fine weather, or close to a window from which the clouds to be observed can be seen. The instrument must be properly oriented, so that the four points marked N., E., S., and W. on the frame shall correspond to the four chief compass directions. The zero (0°) of the movable brass scale is usually put at the S. Hence, if a cloud is found moving from exactly SW., the angular measurement of its direction of motion will be 45°. If a cloud is moving from due E., the angular measurement of its direction of motion will be 270°.

When the sky is completely overcast with a uniform layer of cloud, it is usually impossible to determine any direction of movement, because of the difficulty of selecting and keeping in view, on the mirror, some particular point of cloud.

Observations with the nephoscope may be made as often as is desired, and should be entered in an appropriate column in the record book.

Tabulation of Observations.—A convenient form of table which may be used in the complete instrumental observations is given on the next page. The number of columns and their arrangement may, of course, be varied to suit the number and the nature of the records.

Summary of Observations.—In the preceding chapter we have seen how to obtain the mean monthly temperature from the daily observations, the frequency of the different wind directions for each month, and the total monthly precipitation. The addition of the new instruments, the maximum and minimum thermometers, the psychrometer, the anemometer, and the nephoscope, enables us to obtain the following additional data in our monthly summaries.

Temperature.—The *mean monthly temperature* may be obtained from the maximum and minimum temperatures as follows: Add together all the daily maximum and minimum temperatures for a month. Divide this sum by the total number of readings you have made of each thermometer (*i.e.*, one reading of the maximum and one of the minimum each day, making two readings a day), and the result will be the *mean monthly temperature* derived from the maximum and minimum temperature. This is a more accurate mean temperature than the one noted in the summary of the preceding chapter.

Add together all the maximum temperatures noted during one month. Divide this sum by the number of observations,

TABLE FOR METEOROLOGICAL RECORD.

	Date.
	Hour.
	Pressure (in inches).
	Dry Bulb.
	Wet Bulb.
	Max.
	Min.
	Dew-Point.
	Relative Humidity.
	Direction.
	Velocity (miles per hr.).
	Total Miles per Day.
	Kind.
	Amount (in tenths).
	Direction of Movement.
	Angular Altitude.
	Time of Beginning and Ending.
	Kind.
	Amount.
	Remarks.

Temperature. — Humidity. — Wind. — Clouds. — Precipitation.

SUMMARY OF OBSERVATIONS.   45

and the result gives the *mean maximum temperature* for the month.

A similar operation applied to the minimum temperatures gives the *mean minimum temperature* for the month.

In meteorological summaries it is customary also to include the *absolute maximum* and the *absolute minimum* temperatures, *i.e.*, the highest and lowest single readings of the thermometer made during each month. These can easily be determined by simple inspection of your record book. Note also the dates on which the absolute maximum and the absolute minimum occurred.

The *absolute monthly range* of temperature is the difference, in degrees, between the absolute maximum and the absolute minimum.

Humidity. — The *mean relative* humidity is obtained by adding together all the different percentages of relative humidity obtained during the month, and dividing this sum by the whole number of observations of this weather element.

Wind. — The *mean velocity* of the wind corresponding to the different wind directions is readily obtained by adding together all the different velocities (in miles per hour) observed in winds from the different directions, and dividing these sums by the number of cases. The wind summaries will thus give the frequency of the different directions during each month, and the corresponding mean velocities.

The *maximum hourly wind velocity* is obtained by inspection of the velocity column.

The *total monthly wind movement* is readily deduced from the daily records in the twelfth column of the table on p. 44.

State of the Sky. — In connection with the more advanced records described in this chapter, the observations of cloudiness should record the number of *tenths* of the sky cloudy, as closely as the amount can be estimated by eye, instead of indicating the state of the sky as *cloudy, fair,* etc. A detailed record of cloudiness in tenths gives opportunity to determine the *mean cloudiness* for each month, by averaging, as in the case of the other means already described.

If nephoscope observations are made, the monthly summary may include the *mean direction of cloud movement* for each month. This is obtained by adding together all the different angular measurements of directions of cloud movement, and dividing by the whole number of such observations.

By means of your monthly summaries compare one month with another. Notice how the means and the extremes of the different weather elements are related; how they vary from month to month. Are there any *progressive* changes in temperature, cloudiness, precipitation, etc., from month to month? What are the changes? Summarize, in a short written statement, the meteorological characteristics of each month as shown by your tables.

PART III. — EXERCISES IN THE CONSTRUCTION OF
WEATHER MAPS.

CHAPTER IV.

THE DAILY WEATHER MAP.

THE first daily weather maps were issued in connection with the Great Exhibition of 1851 in London. The data were collected by the Electric Telegraph Company and transmitted to London over its wires. These maps were published and sold daily (excepting Sundays) from Aug. 8 to Oct. 11, 1851. The first official weather map of the United States Weather Service was prepared in manuscript on Nov. 1, 1870, and on Jan. 14, 1871, the work of manifolding the maps for distribution was begun at Washington. Previous to the publication of this government map, Professor Cleveland Abbe had issued in Cincinnati, with the support of the Chamber of Commerce of that city, the first current weather maps published in the United States (Feb. 24 to Dec. 10, 1870). In France, daily weather maps have been published continuously since Sept. 16, 1863.

Two things are essential for the publication of a daily synoptic weather map ; *first*, simultaneous meteorological observations over an extended area ; and, *second*, the immediate collection of these observations by telegraph. The weather map of the United States is based on simultaneous observations made at about 150 stations in different parts of this country, besides several coöperating stations in Canada, Central America, Mexico, and the West Indies. At each of our stations, whose location may be seen on any weather map, the Weather Bureau employs one or more observers, who, twice a day, at 8 A.M. and 8 P.M., "Eastern Standard Time," make regular observations of the ordinary weather elements, *i.e.*, temperature, pressure, humidity, wind direction and velocity, precipitation, cloudiness, etc. The instruments at these stations are all standard, but the completeness of the equipment varies according to the importance of the station. The 8 A.M. observations are the only ones now generally used in the preparation of weather maps. When the Weather Service was first established, tri-daily charts were for some time issued from the central office in Washington. On April 1, 1888, the number was reduced to two a day, and on Sept. 30, 1895, a further change was made, and now there is but one map a day.

The 8 A.M. observations, as soon as made, are corrected for certain instrumental errors, and the barometer readings are reduced to sea level. The data are then put into cipher, not for secrecy, but to facilitate transmission and to lessen the chances of error, and are telegraphed from all parts of the country to the central office of the Weather Bureau in Washington. Besides sending their own messages to Washington, all the important stations of the Weather Bureau receive, by a carefully devised system of telegraphic circuits, a sufficient number of the reports from other stations to enable their observers to draw and issue local weather maps.

The observations are received at the central office of the Weather Bureau in Washington by special wires, and are usually all there within an hour after the readings were made. As the messages are received in the forecast room, they are translated from the cipher back again into the original form, and the data are entered upon blank maps. The official charged with making the forecasts then draws upon the maps lines of equal temperature, lines of equal pressure, lines of equal pressure-change and temperature-change during the past 24 hours. These several sets of lines, together with those showing the regions of precipitation during the past 24 hours, furnish the necessary data on which the forecasts can be based. In other words, the forecast official has before him, on the several maps, a bird's-eye view of the weather conditions over the United States as they were an hour before, and also of the changes that have taken place in these conditions during the preceding 24 hours. Thus, by knowing the general laws which govern the movements of areas of high and low temperature, of fair and stormy weather, across the country, he can make a prediction as to the probable conditions which any state or section of the country will experience in 12, 24, or 36 hours.

In a later chapter some suggestions will be given for studies of forecasting.

The forecasts made in Washington, and printed on the Washington daily weather map, relate to all sections of the United States, and include predictions of cold waves, killing frosts, storm winds, river floods, and the like, besides the ordinary changes in weather conditions. These forecasts, as soon as made, are at once given to the local newspapers and to the press associations. They are also sent by telegraph to all regular stations of the Weather Bureau, and to all stations at which cautionary or storm signals are to be displayed, along the Atlantic or Gulf coasts, and on the Great Lakes.

The Washington weather map is about 24 by 16 inches in size, and is newly lithographed each day. The total number of maps issued from the central office during the fiscal year ending June 30, 1898, was 310,250. In addition to these, there are now 84 stations of the Weather Bureau in different parts of the country, at which daily weather maps are issued and local forecasts made. These latter forecasts are made by a corps of local forecast officials, each of whom has to make the weather prediction for his own district. At first, and until within a few years, one predicting officer in Washington made all the forecasts for the country, but it was found better to have the country divided into geographical sections, over each one of which the meteorological conditions are fairly similar, and to have a local forecast official in charge of each section. These local forecast officials have the double advantage of being able to study the weather conditions over the whole country, as sent them by telegraph each morning, and also of knowing the special peculiarities of their own regions. This enables them to make more accurate predictions than can be made by an official who may be one or two thousand miles distant, in Washington.

The greater portion of the maps issued at the map stations outside of Washington are prepared by what is known as the chalk-plate process, suggested by Mr. J. W. Smith, local forecast official at Boston. This process is as follows: A thin covering of specially prepared chalk, $\frac{1}{8}$ of an inch in thickness, is spread upon a steel plate of the size of the prospective weather map. On this chalk are engraved, by means of suitable instruments, the various weather symbols, the lines of equal pressure and of equal temperature, and the wind arrows. The plate is then stereotyped in the ordinary way, and printed on a sheet prepared for the purpose, which has a blank outline map of the United States at the top, and space in the lower half for the forecasts, summary, and tables.

The size of the chalk-plate map itself is 10 by $6\frac{1}{2}$ inches; the size of the whole sheet, which includes also the text and tables, 16 by 11 inches. Weather maps prepared by the chalk-plate process are now issued from 28 of the 84 stations which publish daily maps. At the remaining stations the maps are prepared by a stencil process, the size of the map being $13\frac{1}{2}$ by 22 inches. The total number of weather maps issued at the various stations during the fiscal year 1897–1898 was 5,239,300.

Besides recording the usual meteorological data, and publishing weather maps and forecasts, the various stations of the Weather Bureau serve as distributing centers for cold wave, frost, flood, and storm warnings. These warnings are promptly sent out by telegraph, telephone, and mail. Besides these usual methods of distributing forecasts, other means have also been adopted. In some places factory whistles are employed to inform those within hearing as to the coming weather; railway trains are provided with flags, whose various colors announce to those who are near the train fair or stormy weather, rising or falling temperature; and at numerous so-called "display stations," scattered all over the country, the forecasts are widely disseminated by means of flags.

CHAPTER V.

TEMPERATURE.

A. **Lines of Equal Temperature.** — Temperature is the most important of all the weather elements. It is therefore with a study of the distribution of temperature over the United States, and of the manner of representing that distribution, that we begin our exercises in map drawing. In carrying out the work we shall proceed in a way similar to that adopted by the officials of the Weather Bureau in Washington and at the other map-publishing stations over the country.

Enter on a blank weather map the temperature readings found in the first column of the table in Chapter VIII. These readings are given in degrees of the ordinary Fahrenheit scale [those which are preceded by the minus sign (−) being below zero], and were made at the same time (7 A.M., "Eastern Standard Time") all over the United States. Make your figures small but distinct, and place them close to the different stations to which they belong. This is done every morning at the Weather Bureau in Washington, when the telegraphic reports of weather conditions come in from all over the country. When all the temperature readings have been entered on the outline map, you have before you a view of the actual temperature distribution over the United States at 7 A.M., on the first day of the series. Describe the distribution of temperature in general terms, comparing and contrasting the different sections of the country in respect to their temperature conditions. Where are the lowest temperatures? Where are the highest? What was the lowest thermometer reading recorded anywhere on the morning of this day? At what station was this reading made? What was the highest temperature recorded? And at what station was this reading made?

Notice that the warmest districts on the map are in Florida, along the Gulf Coast, and along the coast of California. The marked contrasts in temperature between the Northwest and the Pacific and Gulf Coasts at once suggest a reason why Florida and Southern California are favorite winter resorts. To these favored districts great numbers of people who wish to escape the severe cold of winter in the Northern States travel every year, and here they enjoy mild temperature and prevailingly sunny weather. To the cold Northwest, on the other hand, far from the warm waters of the Pacific, where the days are short and the sun stands low in the sky, no seekers after health travel. This annual winter migration from the cities of the North to Florida and Southern California has led to the building of great hotels in favored locations in these States, and during the winter and spring fast express trains, splendidly equipped, are run from north to south and from south to north along the Atlantic Coast to accommodate the great numbers of travelers between New York, Philadelphia, Boston, Chicago, and other large northern cities, and the Florida winter resorts. Southern California also is rapidly developing as a winter resort, and rivals the far-famed Riviera of Southern Europe as a mild and sunny retreat from the severe climates of the more northern latitudes. The control which meteorological conditions exercise over travel and over habitability is thus clearly shown. Florida and Southern California are also regions in which, owing to the mildness of their winter climates, certain fruits, such as oranges and lemons, which are not found elsewhere

TEMPERATURE. 53

in the country, can be grown out of doors, and these are shipped to all parts of the United States.

Let us take another step in order to emphasize more clearly the distribution of temperature over the United States on the first day of our series. Draw a line which shall separate all places having a temperature *above* 30° from those having temperatures *below* 30°, 30° being nearly the freezing point and, therefore, a critical temperature. Evidently this will help us to make our description of the temperature distribution more detailed. If this line is to separate places having temperatures *above* 30° from those having temperatures *below* 30°, it must evidently pass through all places whose temperature is exactly 30°. Examine the thermometer readings entered on your map to see whether there are any which indicate exactly 30°. You will find this reading at Norfolk, Va., Wilmington, S. C., Atlanta, Ga., Chattanooga, Tenn., Ft. Smith, Ark., and Portland, Ore. Through all these stations the line of 30° must be drawn. Begin the line on the Atlantic Coast at Norfolk, Va., and draw it wherever you find a thermometer reading of 30°. It is best to trace the line faintly with pencil at first, so that any mistakes can be easily rectified, and it should be drawn in smooth curves, not in angles. From Norfolk the line must run southwest through Wilmington, and then westward through Atlanta, passing just north of Augusta, which has 31°. From Atlanta the line goes northwest through Chattanooga, and thence westward, curving south of Memphis (28°) and Little Rock (26°), and then northwestward again through Ft. Smith.

In fixing the *exact* position of the 30° line south of Memphis and Little Rock, the following considerations must be our guide: Memphis has 28°; Vicksburg has 35°. Neither of these stations has 30°. Suppose, however, that you had started from Memphis, with a thermometer, and had traveled very rapidly to Vicksburg. The thermometer reading at starting in Memphis would have been 28°, and at the end of your journey in Vicksburg it would have been 35°, presuming that no change in temperature at either station took place during the journey. Evidently the mercury rose during the journey, and in rising from 28° to 35° it must, somewhere on the way, have stood at exactly 30°. Now this place, where the temperature was exactly 30°, is the point through which our 30° line ought to pass. How are we to determine its location? Assume, as is always done in such cases, that the temperature increased at a uniform rate between Memphis and Vicksburg. The total rise was from 28° to 35° = 7°. In order to find a temperature 7° higher than at Memphis, you had to travel the whole distance from Memphis to Vicksburg. Suppose you had only wished to find a temperature 5° higher. Then, assuming a uniform rate of increase between the two stations, you would have had to travel only $\frac{5}{7}$ of the distance, and your thermometer at that place would have read 28° + 5° = 33°. But assume you had wanted to find the place where the thermometer stood at 30°. In this case you would have been obliged to go but $\frac{2}{7}$ of the total distance from Memphis to Vicksburg, and at that point your thermometer reading would have been 28° + 2° = 30°, which is the point we wish to find. In this way, then, when we do not find the *exact* temperature we are looking for on the map, we can calculate where that temperature prevails by noting places which have temperatures somewhat higher and somewhat lower, and proceeding as in the case just described. Take another example. Little Rock, Ark., has 26°; Shreveport, La., has 40°. 40° − 26° = 14°, which is the total difference. From 26° to 30° is 4°. Therefore a point $\frac{4}{14}$ or $\frac{2}{7}$ of the distance from Little Rock to Shreveport should have a temperature of 26° + 4° = 30°, which is the point we wish to find, and through which our 30° line must pass.

From Ft. Smith the line cannot go north or northwest or west, because the temperatures there are all below 30°. To the south the temperatures are all above 30°. Evidently there is

TEMPERATURE. 55

only one direction in which you can prolong the line, and that is to the southwest. Temperatures of 30° cannot be found north of El Paso (28°), because there the temperature distinctly falls, Santa Fé having 4°, Denver, − 14°, and Cheyenne, − 23°. Therefore temperatures *above* 28° must be found south of El Paso. From Ft. Smith you may, therefore, continue the 30° line southwest and west, passing close to El Paso, but to the south of it. In determining the further course of the 30° line, note that Yuma and all the California stations have temperatures above 30°, while Winnemucca, Nev., has 13°, and Portland, Ore., has exactly 30°. From El Paso you may, therefore, continue the line to the northwest, passing up through Central California parallel with the coast line, and to the east of all the California stations and of Roseburg, Ore., and thence running through Portland, Ore., ending just west of Seattle, Wash. Notice that the 30° line should be nearer to Sacramento, Cal., with 36°, than to Red Bluff with 44°.

Thus you have drawn the line which passes through all places that have a temperature of 30° on the map under discussion. This may be called *a line of equal temperature*. *Isotherm*, a compound of two Greek words meaning *equal temperature*, is the name given in meteorology to such lines as this. You have drawn the isotherm of 30°. All parts of the United States north and east of this line are below 30°, while all districts south and west of it are above 30°. You see, therefore, how much easier the drawing of this one line has made the description of the temperature distribution over the United States.

Carry this process a step further by drawing the line which shall pass through all places with a temperature of 40°. This line begins at Jacksonville, Fla. (40°), and runs west, passing between Montgomery, Ala. (33°), and Pensacola, Fla. (46°). Thence it turns to the northwest, passing between Vicksburg, Miss. (35°), and New Orleans, La. (48°), and through Shreveport, La. (40°). From Shreveport it turns to the southwest, passing to the north and west of Palestine, Tex. (46°), and down through San Antonio, Tex. (40°). Its further exact location cannot be determined in Mexico, because there are no observations from Mexican stations, but the readings at Yuma, Ariz. (41°), and at San Diego (42°), Los Angeles (44°), San Francisco (45°), Red Bluff (44°), and Cape Mendocino (43°), all in California, show that the 40° isotherm may be started again just north of Yuma, and may be carried up through California, nearly parallel with the Pacific Coast, ending between Cape Mendocino, Cal. (43°), and Roseburg, Ore. (37°). You have now drawn the isotherms of 30° and of 40°, and in order to avoid confusion, mark the ends of the first line 30° and the ends of the second line 40°.

Isotherms on weather maps are drawn for every even 10° of temperature. They are drawn in smooth curves and not in angular sections. Two isotherms cannot cross one another, for if they did you would have two temperatures, differing by 10°, at the point of crossing, which is obviously impossible. Complete the chart for this day by drawing the remaining isotherms, *i.e.*, those for 50°, 20°, 10°, 0°, − 10°, − 20°, and − 30°, bearing in mind what has been said in regard to the determination of the positions of isotherms when the *exact* temperature you are seeking is not given on the map.

The dotted lines in Fig. 18 show the positions of the isotherms when drawn. Notice how clearly the temperature distribution now stands out, and how simple the description of that distribution has become. Observe that the isotherms, although more or less irregular, show a good deal of uniformity in their general courses, and this uniformity is a great assistance in drawing them. Study the distribution of temperature on this map, and the positions of the isotherms, very carefully.

Construct isothermal charts for the remaining days of the series. Use a new blank map for each day, and take the temperature observations from the table in Chapter VIII. Proceed as in the case of the first day. Draw the isotherms for every even 10° of temperature, taking care to study the course of each line

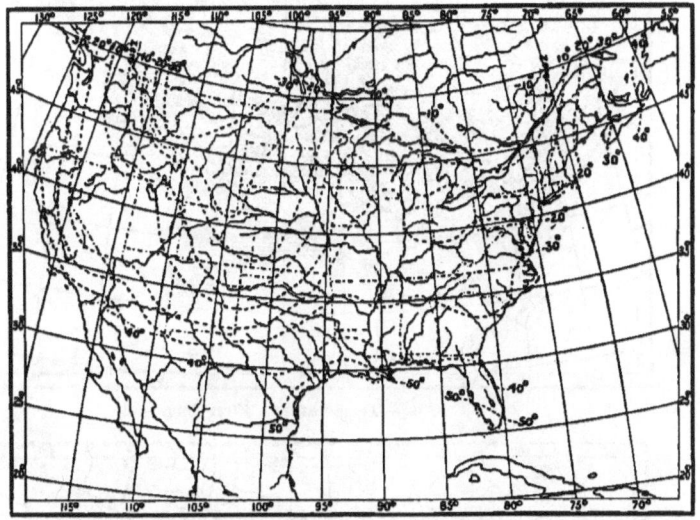

FIG. 18.—Isotherms. First day.

before you begin to draw the line. The charts when completed form a series in which the temperature distribution over the United States is shown at successive intervals of 24 hours.

In order to bring out the temperature distribution on the maps more clearly, color (with colored pencils or water colors) all that portion of each map which lies within the $-20°$ isotherm a dark blue; that portion which is between the 0° isotherm and the $-20°$ isotherm a somewhat lighter shade of blue, and those districts which are between 0° and $+30°$ a still lighter blue. The portion of the map above 30° and below 40° may be left uncolored, while the districts having temperatures over 40° may be colored red. In the map for the third day the district which has temperatures below $-50°$ should be colored darker blue than any shade used on the other maps, or black, in order to

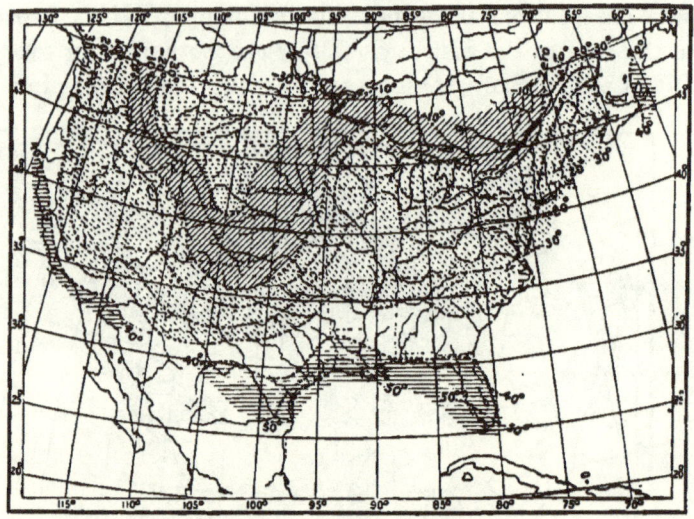

FIG. 19.—Temperature. First Day.

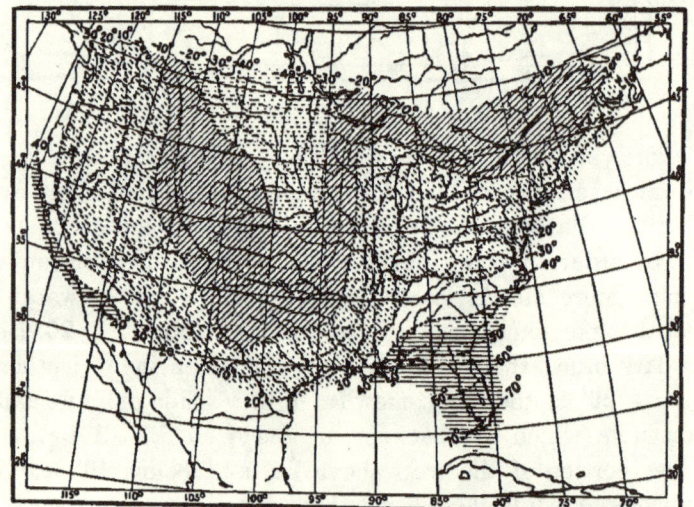

FIG. 20.—Temperature. Second Day.

emphasize the extremely low temperatures there found. Figs. 19–24, on which the isotherms are shown, also illustrate the

TEMPERATURE. 59

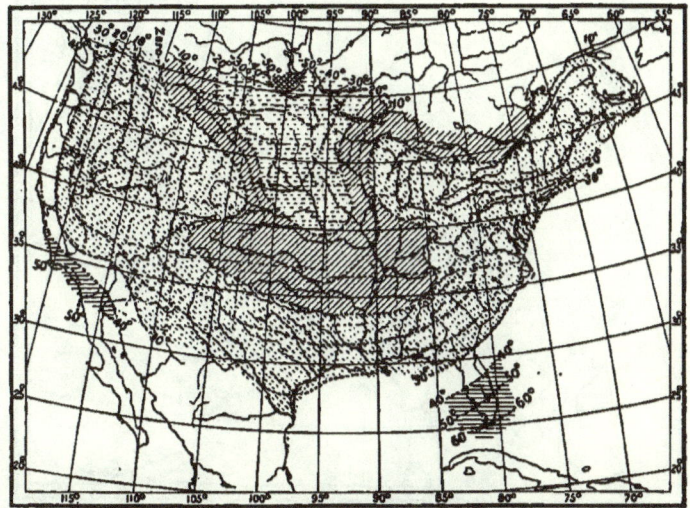

FIG. 21.—Temperature. Third Day.

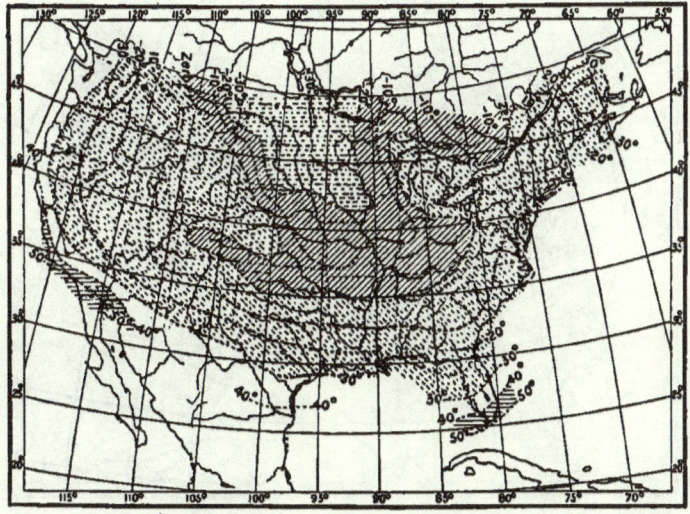

FIG. 22.—Temperature. Fourth Day.

appearance of these maps when the different temperature areas are colored, as has been suggested.

60   CONSTRUCTION OF WEATHER MAPS.

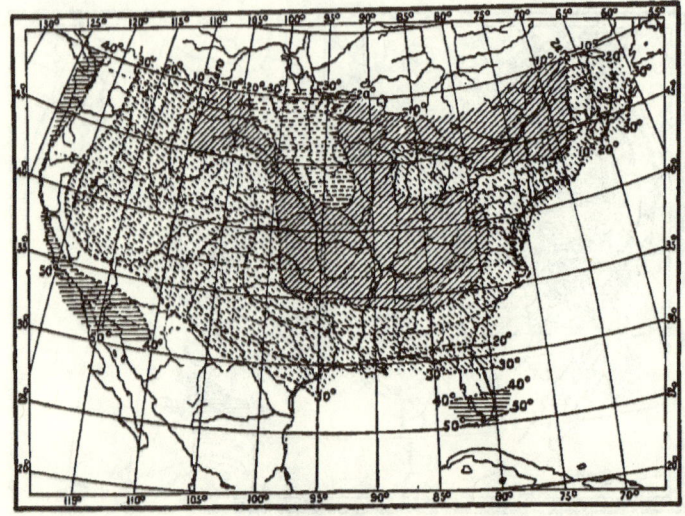

FIG. 23. — Temperature. Fifth Day.

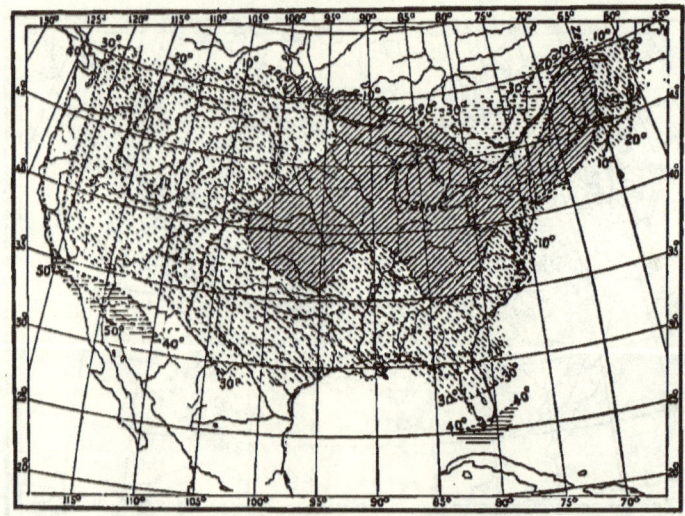

FIG. 24. — Temperature. Sixth Day.

Study the maps individually at first. Describe the temperature distribution on each map. Ask yourself the following questions in each case: Where is it coldest? Where warmest? What is the lowest temperature on the map? What is the highest? At what stations were these readings made?

Then compare the successive maps and answer these questions: What changes have taken place in the intervening 24 hours? In what districts has the temperature risen? What is the greatest rise that has occurred? Where? In what districts has the temperature fallen? What was the greatest fall in temperature and where did it occur? Has the temperature remained nearly stationary in any districts? In which? You will find it a help in answering such questions to make out a table of all the stations, and to indicate in columns, after the names of the stations, the number of degrees of rise or fall in temperature at each place during the 24-hour interval between the successive maps. When the temperature is higher at any station than it was on the preceding day, note this by writing a plus sign (+) before the number of degrees of rise in temperature. When the temperature has fallen, put a minus sign (−) before the number of degrees of fall. Thus, New Orleans, La., had a temperature of 48° on the first day. On the second it had 33°. Therefore the change at New Orleans was − 15° in the 24 hours. At Key West, Fla., the change was + 11° in the same time.

Write a brief account of the temperature distribution on each day of the series, and of the changes which took place between that day and the one preceding, naming the districts and States over which the most marked falls and rises in temperature occurred, with some indication of the amount of these changes. Note especially the changes in position, and the extent, of the districts with temperatures below − 20°; between 0° and −20°, and between 30° and 0°. Write out a clear, concise statement of the temperature distribution and changes shown on the whole set of six maps.

Cold Waves. — The series of charts for these six days furnishes an excellent illustration of a severe cold wave.

A *cold wave*, as the term is now used by the Weather Bureau, means, during December, January, and February, a fall in temperature of from 20° to 16° in 24 hours, with a resulting reduction of temperature to between 0° and 32°, and, during the months from March to November inclusive, a fall of from 20° to 16° in 24 hours, with a reduction of temperature to from 16° to 36°. During December, January, and February a *cold wave* means the following falls and reductions of temperature. Over the Northwestern States, from western Wisconsin to Montana, including Wyoming, Nebraska, and western Iowa, and over northeastern New York and northern New Hampshire, northern Vermont and northern Maine, a fall of 20° or more to zero or below; over southern New England and adjoining districts, the Lake region, the central valleys and west to Colorado, including northern New Mexico and northwestern Texas, a fall of 20° or more to 10° or below; over southern New Jersey, Delaware, eastern Maryland, Virginia, western North Carolina, northwestern South Carolina, northern Georgia, northern Alabama, northern Mississippi, Tennessee, southern Kentucky, Arkansas, Oklahoma, and southern New Mexico, a fall of 20° or more to 20° or below; over eastern North Carolina, central South Carolina, central Georgia, central Alabama, central Mississippi, central and northern Louisiana and central and interior Texas, a fall of 18° or more to 25° or below; along the Gulf coasts of Texas, Louisiana, Mississippi, and Alabama, over all of Florida, and over the coasts of Georgia and South Carolina, a fall of 16° or more to 32° or below: From March to November inclusive a *cold wave* means falls of temperature of the same amounts over the same districts, with resulting temperatures of 16°, 24°, 28°, 32°, and 36° respectively.

Notice that the region from which the greatest cold came in this cold wave is Canada. In that northern country, with its short days and little sunshine, and its long, cold nights, everything is favorable to the production of very low temperatures.

Cold waves occur only in winter. In the summer cool spells, with similar characteristics, may be called *cool waves*.

Cold-Wave Forecasts. — A severe cold wave in winter does much damage to fruit and crops growing out of doors in our Southern States, and to perishable food products in cars, on the way from the South to supply the great cities of the North. Therefore it is important that warnings should be issued giving early information of the coming cold, so that farmers and fruit growers and shippers may take every precaution to protect their crops and produce. Our Weather Bureau takes special pains to study the movements of cold waves and to make forecasts of them, and so well are the warnings distributed over the country that the fruit growers and the transportation companies, and the dealers in farm produce, are able every winter to save thousands of dollars' worth of fruit and vegetables which would otherwise be lost. Cold-wave warnings are heeded by many persons besides those who are directly interested in fruits and farm products. The ranchmen in the West, with thousands of cattle under their charge ; the trainmen in charge of cattle trains ; the engineers of large buildings, such as hotels, stores, and office buildings, who must have their fires hotter in cold weather, — these and many more watch, and are governed by, the cold-wave forecasts of our Weather Bureau.

Mean Annual and Mean Monthly Isothermal Charts. — We have thus far considered isothermal charts for the United States only, based on the temperature observations made at a single moment of time. It is, of course, possible to draw isothermal charts, the data for which are not the temperatures at a given moment, but are the mean or average temperatures for a month or a year. Such charts have been constructed for other countries besides our own, as well as for the whole world. An isothermal chart based on the mean annual temperatures is known as a *mean annual isothermal chart*. These charts show at once the average distribution of temperature for the month or for the year, just as the ones we have drawn show the distribution of temperature over the United States at a single moment.

B. **Direction and Rate of Temperature Decrease. Temperature Gradient.**—Take your isothermal map for the first day and imagine yourself at Kansas City, Mo. In what direction must you go from Kansas City in order to enter most rapidly into colder weather? In what direction must you go from Kansas City in order to enter most rapidly into warmer weather? Take the case of Salt Lake City. In what direction must you go from that station in order to enter most rapidly into colder weather? Into warmer weather? What are the corresponding directions in the case of Spokane, Wash.? Of Bismarck, N. Dak.? Of Buffalo, N. Y.? Of Montreal, Que.? Of Portland, Me.? Of Sacramento, Cal.?

Draw a line from Kansas City to the nearest point at which there is a temperature 10° lower than at Kansas City. Evidently this point is on the isotherm of 0°, and will be found if a line be drawn from Kansas City towards, and at right angles to, the isotherm of 0°. Continue the line beyond the 0° isotherm in the direction of still lower temperatures, *i.e.*, to the isotherms of − 10°, − 20°, and − 30°. Beyond the isotherm of − 30° the line must stop. Draw similar lines from Seattle, Wash.; Salt Lake City, Utah; Denver, Col.; St. Paul, Minn.; Cleveland, O.; and New York, N. Y. Prolong these lines all across the map, so that they will extend from the regions of highest temperature to those of the lowest. A number of intermediate lines may also be added. Note that the various directions followed by these lines are square to, or at right angles to, the successive isotherms, and that although the lines all run from higher to lower temperatures, they do not all trend in the same direction. These lines may be called *lines of decrease of temperature*. Fig. 25 shows a few of these lines of decrease of temperature drawn for the first day.

Draw similar lines on the other isothermal charts, for the same stations. Are the directions of temperature decrease the same on these charts as on the chart for the first day, for Kansas

City, Seattle, Salt Lake City, Denver, St. Paul, Cleveland, New York? Draw lines of decrease of temperature from the following additional stations: Key West, Fla.; New Orleans, La.; Charleston, S. C.; El Paso, Tex.; San Diego, Cal.; Hatteras, N. C.

Compare the directions of these lines on the different days. How do they change from one day to the next?

Next select some line of decrease of temperature on the map for the first day which begins in Texas, and follow it northward.

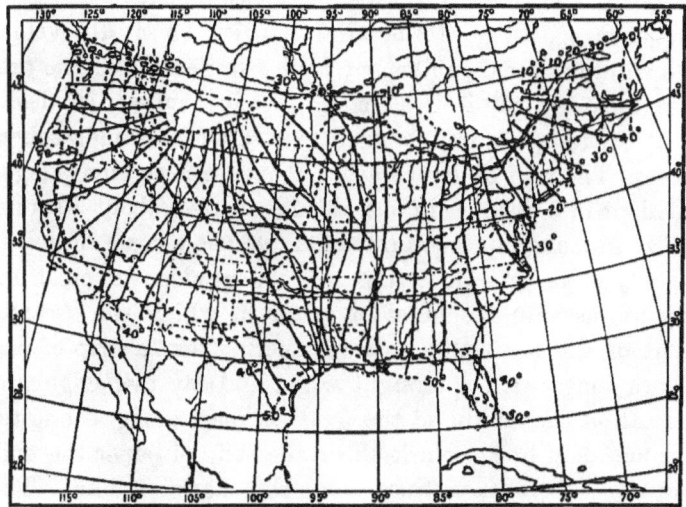

FIG. 25.—Temperature Gradients. First Day.

Where, along this line, is the decrease of temperature most rapid? Evidently this must be where the isotherms are closest together, because every isotherm that is crossed means a change of temperature of 10°, and the more isotherms there are in a given distance, the more rapidly the temperature is changing. Where the isotherms are closest together, a given decrease of temperature is passed over in the least distance, or, conversely, a greater decrease of temperature is experienced in a given distance. Study this question of rapidity or slowness of temperature decrease on the whole series of charts. On which of the charts, and where, do you find the most rapid decrease? The slowest decrease? Is there any regularity in these *rates* of temperature decrease either on one map or in the whole series of maps?

The term *temperature gradient* is used by meteorologists to describe the *direction* and *rate of temperature decrease* which we have been studying.

If we are to compare these rates of temperature change, we must have some definite scale of measurement. Thus, for example, in speaking of the wind velocity we say the velocity of the wind is so many miles per hour; in describing the grade of a railroad we say it is so many feet in a mile. In dealing with these temperature changes, we adopt a similar scheme. We say: The rate of temperature decrease is so many degrees Fahrenheit in a distance of one latitude degree (about 70 miles). In order to make our measurements, we use a scale of *latitude degrees*, just as, in calculating railroad grades, we must have a way to measure the miles of track in which the ascent or descent of the roadbed is so many feet. Take a strip of paper 6 inches long, with a straight edge, and lay this edge north and south at the middle of the weather map, along a longitudinal or meridian line. Mark off on the strip of paper the points where any two latitude lines cross the meridian line. These latitude lines are five (latitude) degrees apart. Therefore divide the space between them on your paper into five divisions, and each of these will measure just one latitude degree. Continue making divisions of the same size until you have ten altogether on the strip of paper. Select, on any weather map, some station lying between two isotherms at which you wish to measure the rate of temperature decrease. Take, for instance, Buffalo, N. Y., on the first day. What you want to find is this: What is the *rate of temperature decrease*, or the *temperature gradient*, at Buffalo? Lay your paper scale of latitude degrees through Buffalo, from the isotherm of 10° to the isotherm of 0°, and as nearly as possible at right angles to the isotherms.[1] Count the number of latitude degrees on your scale between the isotherms of 10° and 0°, on a line running through Buffalo. There are, roughly, about two degrees of latitude in this distance. That is, in the district in which Buffalo lies, the temperature is changing *at the rate* of 10° Fahrenheit (between isotherms 10° and 0°) in two latitude degrees. As our standard of measurement is the amount of change of temperature in one latitude degree, we divide the 10 (the number of degrees of temperature) by the 2 (the number of degrees of latitude), which gives us 5 as the rate of decrease of temperature per latitude degree at Buffalo, N. Y., at 7 A.M., on the first day of the series. The temperature gradient at Buffalo is therefore 5. The rule may be stated as follows : Select the station for which you wish to know the rate of temperature decrease or temperature gradient. Lay a scale of latitude degrees through the station, and as nearly as possible at right angles to the adjacent isotherms. If the station is exactly on an isotherm then measure the distance *from* the station to the nearest isotherm indicating a temperature 10° lower. The scale must, however, be laid perpendicularly to the isotherm, as before. Divide the number of degrees of difference of temperature between the isotherms (always 10°) by the distance (in latitude degrees) between the isotherms, and the quotient is the *rate of temperature decrease per latitude degree*. Or, to formulate the operation :

$$R = \frac{T}{D},$$

in which R = rate ; T = temperature difference between isotherms (always 10°), and D = distance between isotherms in latitude degrees. Thus, a distance of 10 latitude degrees gives a rate of 1 ; a distance of 5 gives a rate of 2 ; a dis-

[1] Unless the isotherms are exactly parallel, the scale cannot be at right angles to both of them. It should, however, be placed as nearly as possible in that position.

tance of 2 gives a rate of 5; a distance of 4 gives a rate of 2.5, etc.

Determine the rates of temperature decrease in the following cases: —

A. For a considerable number of stations in different parts of the same map, as for each of the six days of the series.

And, using the school file of weather maps,

B. For one station during a winter month and during a summer month, measuring the rate on each map throughout the month and obtaining an average rate for the month.

C. For a station on the Pacific Coast, and one on the Atlantic Coast during the same months.

D. For a station on the Gulf of Mexico, or in Florida, and one in the Northwest during a winter month.

E. For a station in the central United States, and one on the Pacific Coast, the Gulf Coast, and the Atlantic Coast, respectively, during different months of the winter and summer.

The determination of the rates of temperature decrease under these different conditions over the United States prepares us for an appreciation of the larger facts, of a similar kind, to be found on the mean annual and mean monthly isothermal charts of various countries, and also of the whole world.

Temperature Gradients on Isothermal Charts of the Globe. — The mean annual isothermal charts of the globe (see page 63) bring out some very marked contrasts in rates of temperature decrease. Thus, along the eastern side of the North American continent the isotherms are crowded close together, while on the western coast of Europe they are spread far apart. Between southern Florida and Maine there is the same change in mean annual temperature as is found between the Atlantic coast of the Sahara and central England. The latter is a considerably longer distance, and this means that the decrease of temperature is much slower on the European Atlantic coast than on the North American Atlantic coast. In fact, the rate of temperature decrease with latitude in the latter case is the most rapid anywhere in the world, in the same distance. These great contrasts in temperature which occur within short distances along the eastern coast of North America have had great influence upon the development of this region, ~~as has been pointed out by Woeikof, an eminent Russian meteorologist.~~ The products of the tropics and of the Arctic are here brought very near together; and at the same time intercommunication between these two regions of widely differing climates is very easy. Labrador is climatically an Arctic land, and man is there forced to seek his food chiefly in the sea, for nature supplies him with little on shore, while southern Florida is quite tropical in its temperature conditions and in the abundance of its vegetation. Between the Pacific coasts of Asia and of North America there is a similar but less pronounced contrast, the isotherms being crowded together on the eastern coast of China and Siberia, and being spread apart as they cross the Pacific Ocean and reach our Pacific Coast.

In general, we naturally expect to find that the temperature decreases as one goes poleward from the equator; from lower latitudes, where the sun is always high in the heavens, to higher latitudes, where it is near the horizon, and its warming effect is less. But there are some curious exceptions to this general rule. The lowest temperatures on the January isothermal chart ($-60°$) are found in northeastern Siberia, and not, so far as our observations go, near the North Pole. If you find yourself at this "cold pole," as it is called, in Siberia in January, you can reach higher temperatures by traveling north, south, east, or west. In other words, here is a case of *increase* of temperature in a *northerly* direction, as well as east, south, and west. Again, there is a district of high temperature ($90°$) over southern Asia in July, from which you can travel south towards the equator and yet reach lower temperatures.

In our winter months the contrasts of temperature in the United States are, as a rule, violent, there being great differences between the cold of the Northwest and the mild air of Florida and the Gulf States. In the summer, on the other hand, the distribution of temperature is relatively equable, the isotherms being, as a rule, far apart. In summer, therefore, we approach the conditions characteristic of the Torrid Zone. These are uniformly high temperatures over large areas. The same thing, on a larger scale, is seen over the whole Northern Hemisphere. During our winter months the isotherms are a good deal closer together than they are during the summer, or, in more technical language, the temperature gradient between the equator and the North Pole is steeper in winter than in summer.

CHAPTER VI.

WINDS.

THE observational work already done, whether non-instrumental or instrumental, has shown that there is a close relation between the *direction of the wind* at any station and the *temperature* at that station. Our second step in weather-map drawing is concerned with the winds on the same series of maps which we have thus far been studying from the point of view of temperature alone.

In the second column of the table in Chapter VIII are given the wind directions and the wind velocities (in miles per hour) recorded at the Weather Bureau stations at 7 A.M., on the first day of the series. Enter on a blank weather map, at each station for which a wind observation is given in the table, a small arrow flying *with* the wind, *i.e.*, pointing in the direction *towards* which the wind is blowing. Make the lengths of the wind arrows roughly proportionate to the velocity of the wind, the winds of higher velocities being distinguished by longer arrows, and those of lower velocities by shorter arrows. The letters *Lt.* (= light) in the table denote wind velocities of 5 miles, or less, per hour.

When you have finished drawing these arrows, you will have before you a picture of the wind directions and velocities all over the United States at the time of the morning observation on this day. (See solid arrows in Fig 26.)

The wind arrows on your map show the wind directions at only a few scattered points as compared with the vast extent of the United States. We must remember that the whole lower portion of the atmosphere is moving, and not merely the

WINDS. 71

winds at these scattered stations. It will help you to get a clearer picture of this actual movement of the atmosphere as a whole, if you draw some additional wind arrows between the stations of observation, but in sympathy with the observed wind directions given in the table and already entered on your map. These new arrows may be drawn in broken lines, and may be curved to accord in direction with the surrounding wind arrows. Heavier or longer lines may be used to indicate faster winds. (See broken arrows, Fig. 26.)

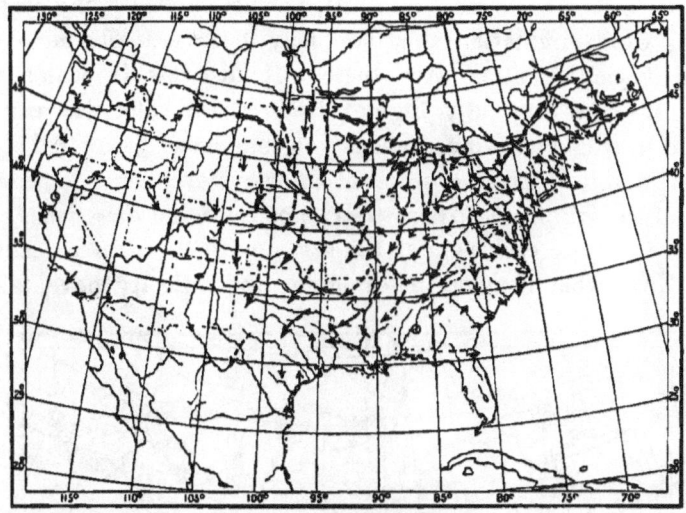

FIG. 26.—Winds. First Day.

It is clear that the general winds must move in broad sweeping paths, changing their directions gradually, rather than in narrow belts, with sudden changes in direction. Therefore long curving arrows give a better picture of the actual drift of the atmospheric currents than do short, straight, disconnected arrows.

Study the winds on this chart with care. Describe the conditions of wind distribution in a general way. Can you discover any apparent relation between the different wind directions in any part of the map? Is there any system whatever in the winds? Write out a brief and concise description of the results of the study of this map.

Enter on five other blank maps the wind directions given in the table in Chapter VIII for the other five days of the series, making, as before, the lengths of the arrows roughly proportionate to the velocity of the wind, and adding extra broken arrows as suggested. (See Figs. 27–31.)

A. Study the whole series of six maps. Describe the wind conditions on each map by itself, noting carefully any system in the wind circulation that you may discover. Examine the wind velocities also. Are there any districts in which the velocities are especially high? Have these velocities any relation to whatever wind systems you may have discovered? If so, include in your description of these systems some consideration of the wind *velocities* as well as of the wind *directions*.

B. Compare each map of the series with the map preceding it. Note what changes in direction and velocity have taken

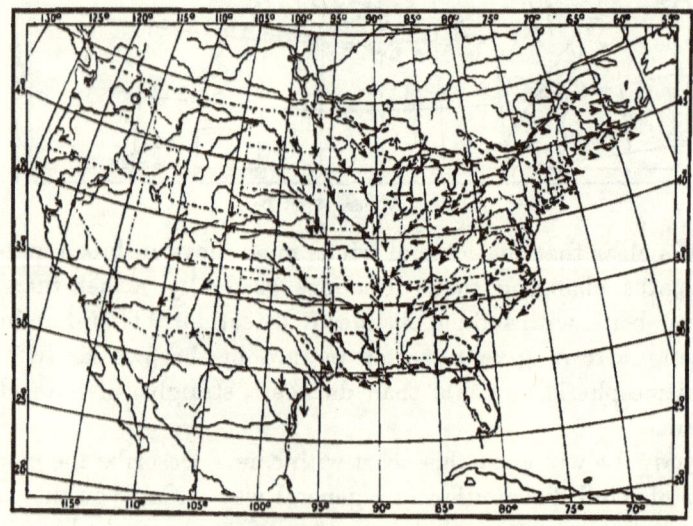

FIG. 27. — Winds. Second Day.

WINDS. 73

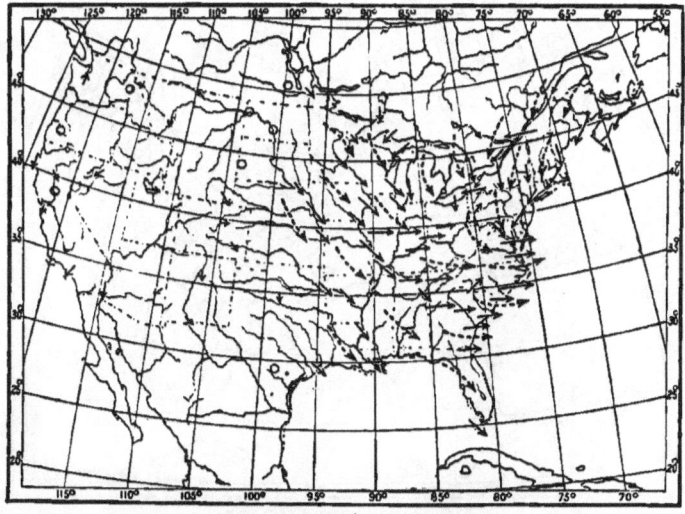

FIG. 28. — Winds. Third Day.

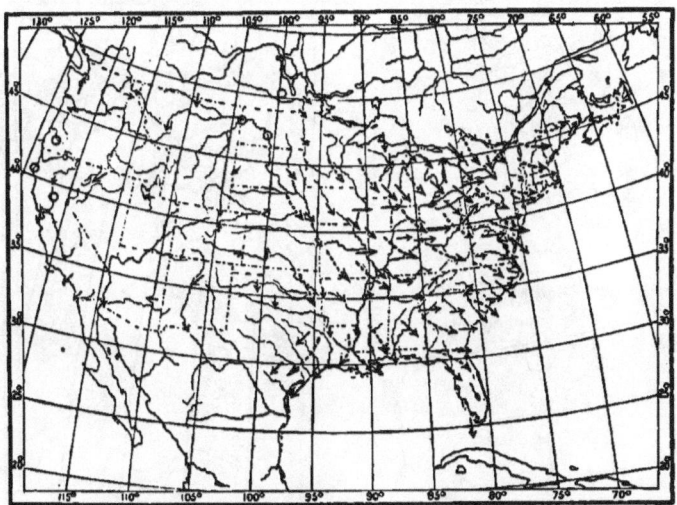

FIG. 29. — Winds. Fourth Day.

place at individual stations. Group these changes as far as possible by the districts over which similar changes have occurred.

74   CONSTRUCTION OF WEATHER MAPS.

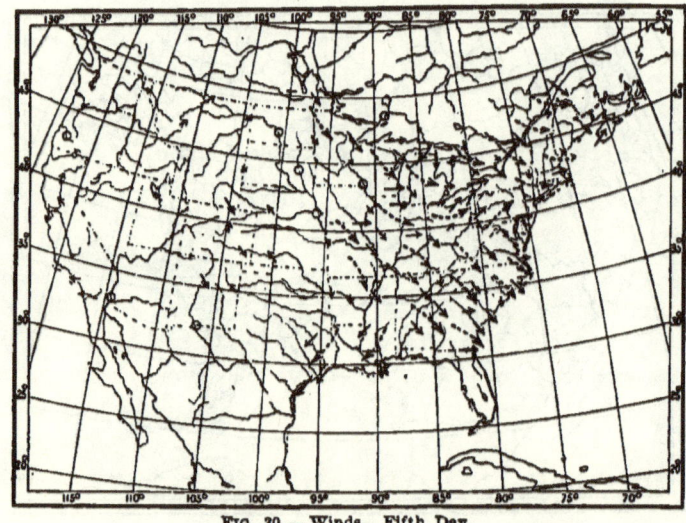

FIG. 30.—Winds. Fifth Day.

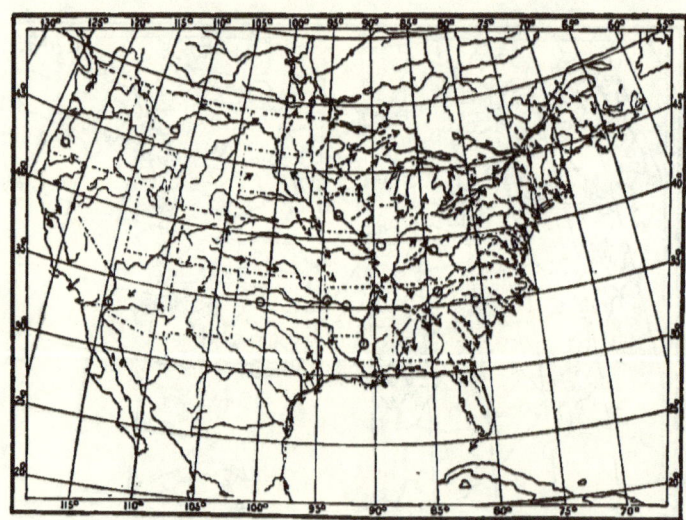

FIG. 31.—Winds. Sixth Day.

Compare the wind systems on each map with those on the map for the preceding day. Has there been any alteration in the position or relation of these systems? Write for each day an account of the conditions on that map, and of the changes that have taken place in the preceding 24-hour interval.

C. Write out a short connected account of the wind conditions and changes illustrated on the whole set of six maps.

In the last chapter we studied the progression of the cold wave of low temperatures in an easterly direction across the United States. Notice now the relation of the winds on the successive maps of our series to the movement of the cold wave. Place your wind charts and isothermal charts for the six days side by side, and study them together. The temperature distribution on the second day differs from that on the first. What are the chief differences? Examine the wind charts for these two days. Do you detect any differences in the wind directions or systems on these days? Do these differences help to explain some of the changes in temperature?

Compare the temperature distribution on the second day with that on the third. What are the most marked changes in the distribution? What changes in the winds on the corresponding wind maps seem to offer an explanation of these variations?

Proceed similarly with each map of the series. Formulate, in writing, the general relation between winds and cold wave, discovered through your study of these charts.

Cold Waves in Other Countries. — Cold waves in the United States come, as has been seen, from the northwest, that being the region of greatest winter cold. In Europe, cold waves come from the northeast. This is because northwest of Europe there is a large body of warm water supplied by the Gulf Stream drift, and therefore this is a source of warmth and not of cold. The cold region of Europe is to the northeast, over Russia and Siberia.

Cold waves have different names in different countries. In southern France the cold wind from the north and northeast is known as the *mistral*, derived from the Latin word *magister*, meaning *master*, on account of its strength and violence. In Russia the name *buran* or *purga* is given to the cold wave when it blows along with it the fine dry snow from the surface of the ground. This *buran* is apt to cause the loss of many lives, both of men and cattle. In the Argentine Republic the coolest wind is from the southwest. It is known as a *pampero*, from the Spanish *pampa*, a *plain*.

Cyclones and Anticyclones. — A system of winds blowing towards a common center (such as is well shown over the Gulf States on the weather map for the second day, and over the middle Atlantic coast on the third day) is called by meteorologists a *cyclone*. The name was first suggested by Piddington early in this century. It is derived from the Greek word for *circle*, and hence it embodies the idea of a circular or spiral movement of the winds. A system of *outflowing* winds, such as that over the northwestern United States shown on the maps for the first five days, and over the western Gulf States on the sixth day, is called an *anticyclone*. This name was proposed by Galton in 1863, and means the opposite of *cyclone*.

CHAPTER VII.

PRESSURE.

A. **Lines of Equal Pressure: Isobars.** — One of the most important weather elements is the *pressure* of the atmosphere. This has already been briefly discussed in the sections on the mercurial barometer (Chapter II). It was there learned that atmospheric pressure is measured by the number of inches of mercury which the weight of the air will hold up in the glass tube of the barometer. Our sensation of heat or cold gives us some general idea as to the air temperature. We can tell the wind direction when we know the points of the compass, and can roughly estimate its velocity. No instrumental aid is necessary to enable us to decide whether a day is clear, fair or cloudy, or whether it is raining or snowing. Unlike the temperature, the wind, or the weather, the pressure cannot be determined by our own senses without instrumental aid. The next weather element that we shall study is pressure.

Proceed as in the case of the thermometer readings. Enter

PRESSURE. 77

upon a blank map the barometer readings for the different stations given in the third column of the table in Chapter VIII. When this has been done you have before you the actual pressure distribution over the United States at 7 A.M., on the first day of the series. Describe the distribution of pressure in general terms. Where is the pressure highest? Where lowest? What are the highest and the lowest readings of the barometer noted on the map? What is the difference (in inches and hundredths) between these readings?

Draw lines of equal pressure, following the same principles as were adopted in the case of the isotherms. The latter were drawn for every even 10° of temperature. The former are to be drawn for every even .10 inch of pressure. Every station which has a barometer reading of an even .10 inch will be passed through by some line of equal pressure. Philadelphia, Pa., with 29.90 must be passed through by the 29.90 line; Wilmington, N. C., with 30.00, must have the 30.00 line passing through it, etc. Chicago, with 30.17 inches, must lie between

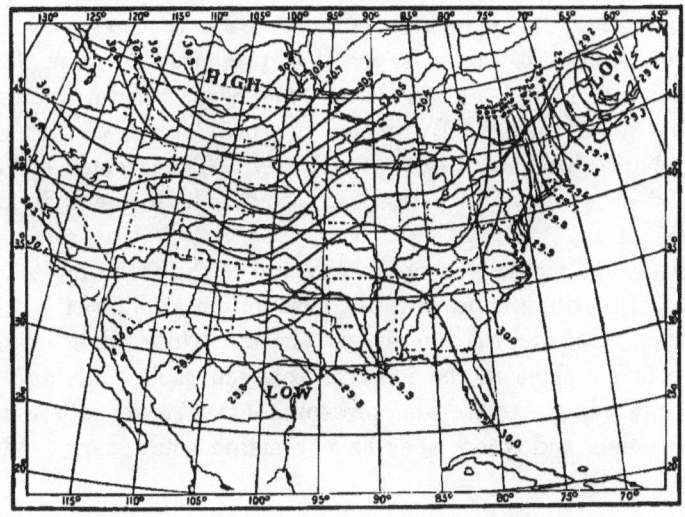

FIG. 32.—Isobars. First Day.

the lines of 30.10 and 30.20 inches, and nearer the latter than the former. Denver, Col., with 30.35 inches, must lie midway between the 30.30 and 30.40 lines (Fig. 32).

Lines of equal pressure are called *isobars*, a word derived from two Greek words meaning *equal pressure*.

Describe the distribution of pressure as shown by the arrangement of the isobars. Note the differences in form between the isotherms and the isobars. The words *high* and *low* are printed on weather maps to mark the regions where the pressure is highest and lowest.

Draw isobars for the other days, using the barometer readings given in the table in Chapter VIII. Figs. 33–38 show the arrangement of the isobars on these days.

The pressure charts may be colored, as indicated by the shading in these figures, in order to bring out more clearly the distribution of pressure, according to the same general scheme as that adopted in the temperature charts. Color *brown* all parts of your six isobaric charts over which the pressures are below 29.50 inches; color *red* all parts with pressure above 30.00 inches. Use a *faint shade of brown* for pressures between 29.50 inches and 29.00 inches, and a *darker shade* for pressures below 29.00 inches. In the case of pressures over 30.00 inches, use a *pale red* for pressures between 30.00 and 30.50 inches, and a *darker shade of red* for pressures above 30.50 inches. By means of these colors the pressure distribution will stand out very clearly. The scheme of color and shading may, of course, be varied to suit the individual fancy.

Study the isobaric chart of each day of the series by itself at first. Describe the pressure distribution on each chart.

Then compare the successive charts. Note what changes have taken place in the interval between each chart and the one preceding; where the pressures have risen; where they have fallen, and where they have remained stationary. Write

PRESSURE. 79

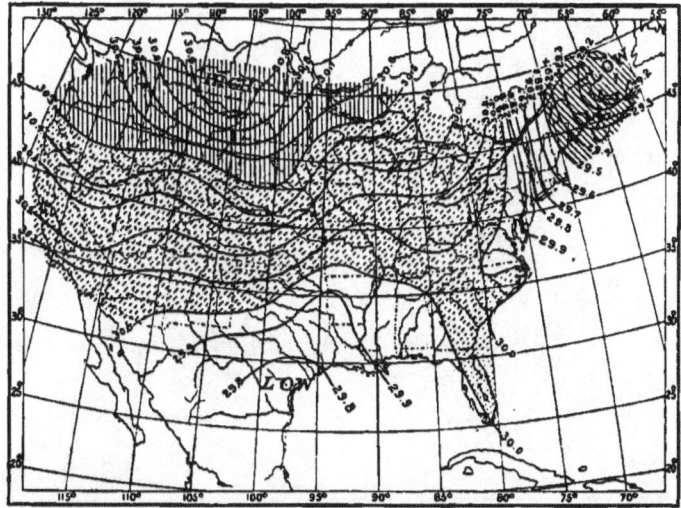

FIG. 33. — Pressure. First Day.

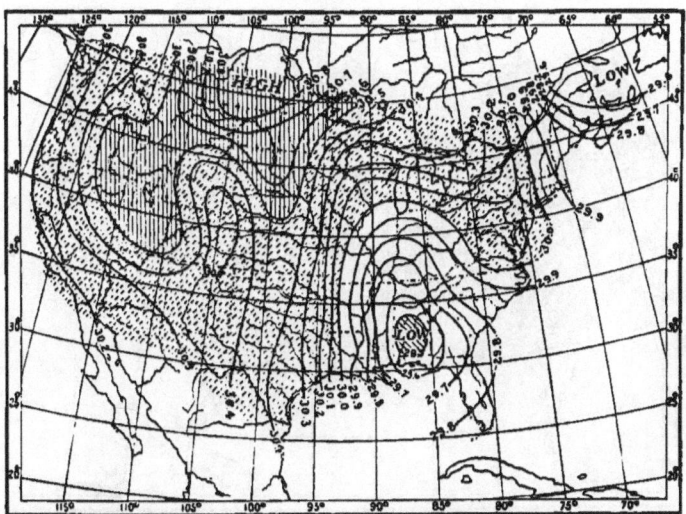

FIG. 34. — Pressure. Second Day.

a brief account of the facts of pressure change illustrated on the whole series of six charts.

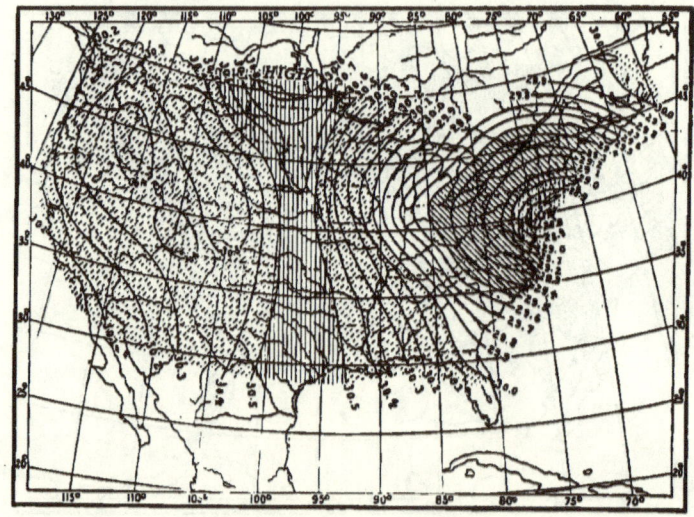

Fig. 35.—Pressure. Third Day.

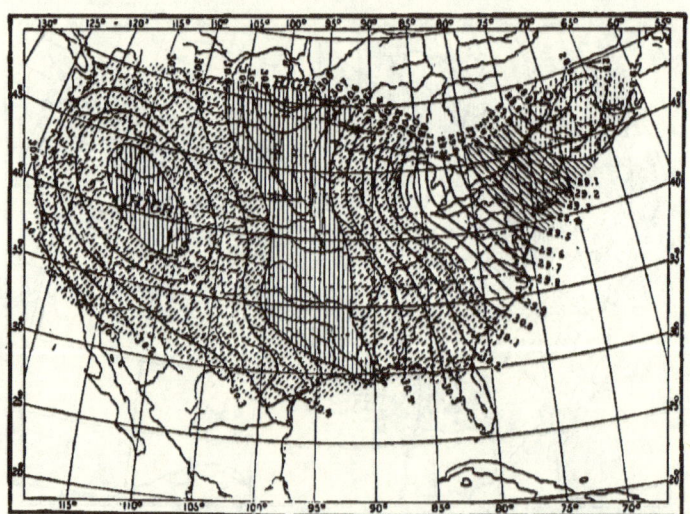

Fig. 36.—Pressure. Fourth Day.

Compare the charts of temperature and of pressure, first individually, then collectively. What relations do you dis-

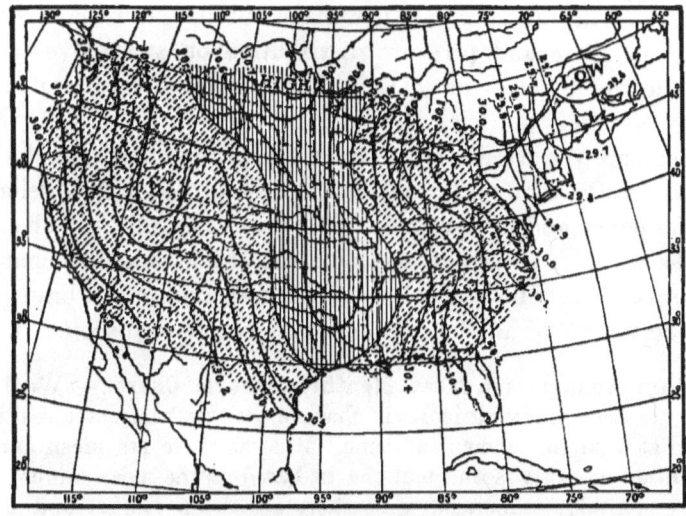

FIG. 37.— Pressure. Fifth Day.

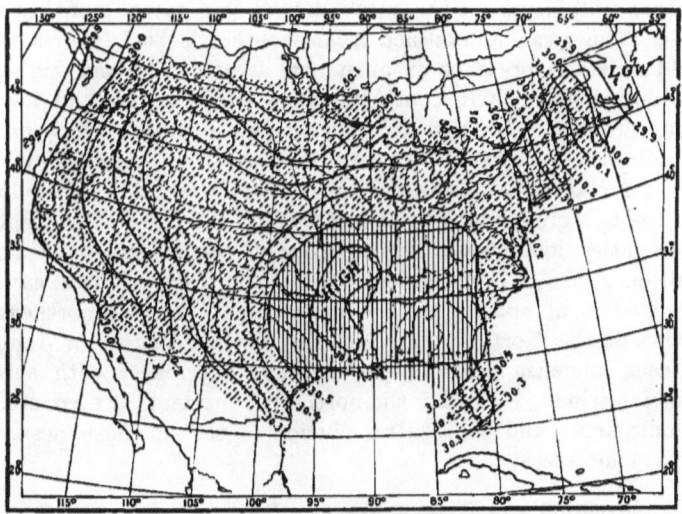

FIG. 38.— Pressure. Sixth Day.

cover between temperature distribution and pressure distribution on the isothermal and the isobaric charts for the same day? What relations can you make out between the changes in temperature and pressure distribution on successive days? On the whole series of maps? Write out the results of your study concisely and clearly.

Compare the wind charts and the pressure charts for the six days. Is there any relation between the direction and velocity of the winds and the pressure? Observe carefully the changes in the winds from day to day on these charts, and the changes in pressure distribution. Formulate and write out a brief general statement of all the relations that you have discovered.

Mean Annual and Mean Monthly Isobaric Charts. — We have thus far been studying isobaric charts based on barometer readings made at a single moment of time. Just as there are mean annual and mean monthly isothermal charts, based on the mean annual and mean monthly temperatures, so there are mean annual and mean monthly isobaric charts for the different countries and for the whole world, based on the mean annual and mean monthly pressures. The mean annual and mean monthly isobaric charts of the world show the presence of great oval areas of low and high pressure covering a whole continent, or a whole ocean, and keeping about the same position for months at a time. Thus, on the isobaric chart showing the mean pressure over the world in January, there are seen immense areas of high pressure (anticyclones) over the two great continental masses of the Northern Hemisphere. These anticyclonic areas, although vastly greater in extent than the small ones seen on the weather maps of the United States, have the same system of spirally *outflowing* winds. Over the northeastern portion of the North Pacific and the North Atlantic, in January, are seen immense areas of low pressure (cyclones) with spirally *inflowing* winds. In July the northern continents are covered by cyclonic areas, and the central portion of the northern oceans by anticyclonic areas.

B. **Direction and Rate of Pressure Decrease : Pressure Gradient.** — In Chapter V we studied the direction and rate of temperature decrease, or temperature gradient. We saw that the direction of this decrease varies in different parts of the map, and that the rate, which depends upon the closeness of the isotherms, also varies. An understanding of temperature gradients makes it easy to study the directions and rates of pressure decrease, or *pressure gradients*, as they are commonly called. Examine the series of isobaric charts to see how the lines of pressure decrease run. Draw lines of pressure decrease for the six isobaric charts, as you have already done on the isothermal charts. When the isobars are near together, the lines of pressure decrease may be drawn heavier, to indicate a more rapid rate of decrease

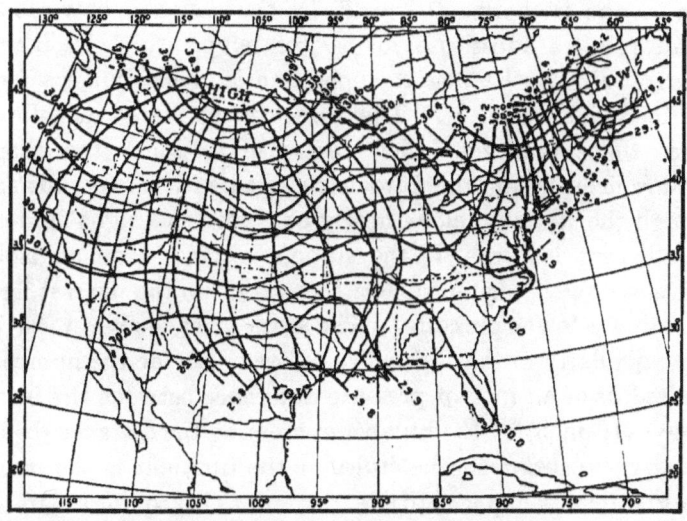

FIG. 39. — Pressure Gradients. First Day.

of pressure. Fig. 39 shows lines of pressure decrease for the first day. Note how the arrangement and direction of these lines change from one map to the next. Compare these lines with the lines of temperature decrease.

Next study the *rate* of pressure decrease. This rate depends upon the closeness of the isobars, just as the rate of temperature decrease depends upon the closeness of the isotherms. Examine the rates of pressure decrease upon the series of isobaric charts. On which charts do you find the most rapid rate?

Where? On which the slowest? Where? Do you discover any relation between rate of pressure decrease and the pressure itself? What relation?

When expressed numerically, the barometric gradient is understood to mean the number of hundredths of an inch of change of pressure in one latitude degree. Prepare a scale of latitude degrees, and measure rates of pressure decrease, just as you have already done in the case of temperature. In this case, instead of dividing the difference in temperature between the isotherms ($10° = T$) by the distance between the isobars (D), we substitute for $10°$ of temperature .10 inch of pressure (P). Otherwise the operation is precisely the same as described in Chapter V. The rule may be stated as follows: Select the station for which you wish to know the rate of pressure decrease or the barometric gradient. Lay your scale through the station, and as nearly as possible at right angles to the adjacent isobars. If the station is exactly on an isobar, then measure the distance *from* the station to the nearest isobar indicating a lower pressure. The scale must, however, be laid perpendicularly to the isobars, as before. Divide the number of hundredths of an inch of pressure difference between the isobars (always .10 inch) by the number expressing the distance (in latitude degrees) between the isobars; the quotient is the rate of pressure decrease per latitude degree. Or, to formulate the operation,

$$R = \frac{P}{D},$$

in which R = rate; P = pressure difference between isobars (always .10 inch), and D = distance between the isobars in latitude degrees.

Determine the rates of pressure decrease in the following cases:—

A. For a number of stations in different parts of the same map, as, *e.g.*, Boston, New York, Washington, Charleston.

New Orleans, St. Louis, St. Paul, Denver, and on the same day.

B. For one station during a winter month and during a summer month, measuring the rate on each map throughout the month, and obtaining an average rate for the month.

Have these gradients at the different stations any relation to the proximity of low or high pressure? To the velocity of the wind?

Pressure Gradients on Isobaric Charts of the Globe. — The change from low pressure to high pressure or *vice versa* with the seasons, already noted as being clearly shown on the isobaric charts of the globe, evidently means that the directions of pressure decrease must also change from season to season. The rates of pressure decrease likewise do not remain the same all over the world throughout the year. If we examine isobaric charts for January and July, we shall find that these gradients are stronger or steeper over the Northern Hemisphere in the former month than in the latter.

CHAPTER VIII.

WEATHER.

HITHERTO nothing has been said about the *weather* itself, as shown on the series of maps we have been studying. By weather, in this connection, we mean the state of the sky, whether it is clear, fair, or cloudy, or whether it is raining or snowing at the time of the observation. While it makes not the slightest difference to our feelings whether the *pressure* is high or low, the *character of the weather* is of great importance.

The character of the weather on each of the days whose temperature, wind, and pressure conditions we have been studying is noted in the table in this chapter. The symbols used by the Weather Bureau to indicate the different kinds of weather on the daily weather maps are as follows: ○ clear; ◐ fair, or partly cloudy; ● cloudy; Ⓡ rain; Ⓢ snow.

Enter on a blank map, at each station, the sign which indicates the weather conditions at that station at 7 A.M., on the first day, as given in the table. When you have completed this, you have before you on the map a bird's-eye view of the weather which prevailed over the United States at the moment of time at which the observations were taken. Describe in general terms the distribution of weather here shown, naming the districts or States over which similar conditions prevail. Following out the general scheme adopted in the case of the temperature and the pressure distribution, separate, by means of a line drawn on your map, the districts over which the weather is prevailingly cloudy from those over which the weather is partly cloudy or clear. In drawing this line, scattering observations which do not harmonize with the prevailing conditions around them may be disregarded, as the object is simply to emphasize the *general* characteristics. Enclose also, by means of another line, the general area over which it was snowing at the time of observation, and shade or color the latter region differently from the cloudy one. Study the weather distribution shown on your chart. What general relation do you discover between the kinds of weather and the temperature, winds, and pressure?

Proceed similarly with the weather on the five remaining days, as noted in the table. Enter the weather symbols for each day on a separate blank map, enclosing and shading or coloring the areas of cloud and of snow as above suggested. In Figs. 40–45 the cloudy areas are indicated by single-line shading, and the snowy areas by double-line shading.

Now study carefully each weather chart with its corresponding temperature, wind, and pressure charts. Note whatever relations you can discover among the various meteorological elements on each day. Then compare the weather conditions on the successive maps. What changes do you note? How are these changes related to the changes of temperature; of wind;

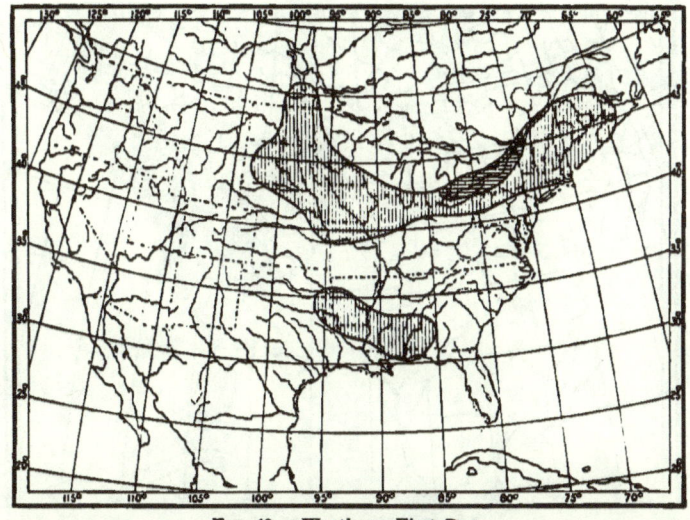

FIG. 40.—Weather. First Day.

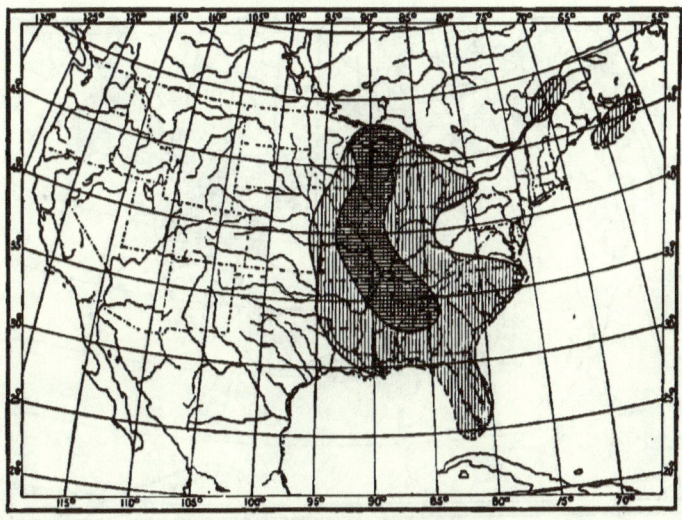

FIG. 41.—Weather. Second Day.

of pressure? Write a summary of the results derived from your study of these four sets of charts.

88   CONSTRUCTION OF WEATHER MAPS.

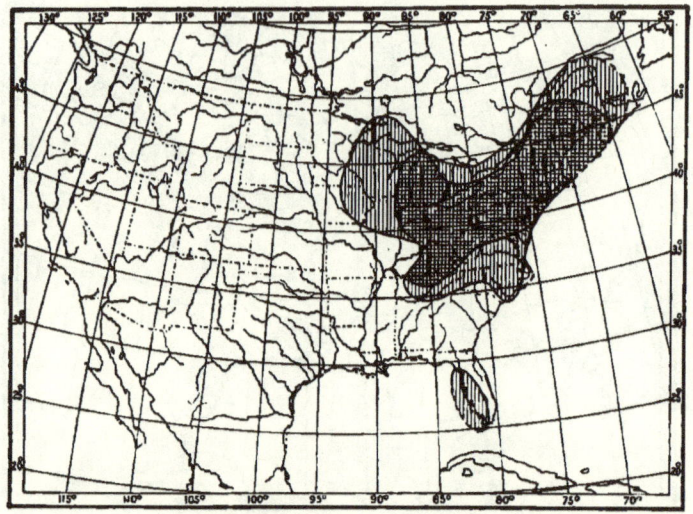

FIG. 42. — Weather. Third Day.

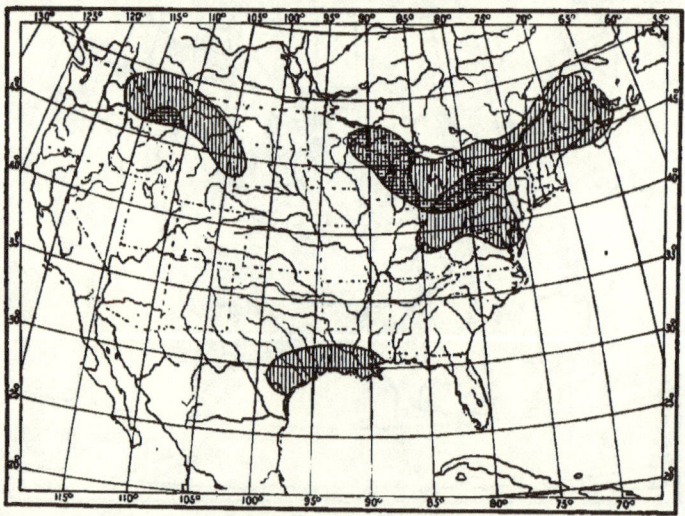

FIG. 43. — Weather. Fourth Day.

The Weather of Temperate and Torrid Zones. — The facts of the presence of clear weather in one region while snow is falling in

WEATHER. 89

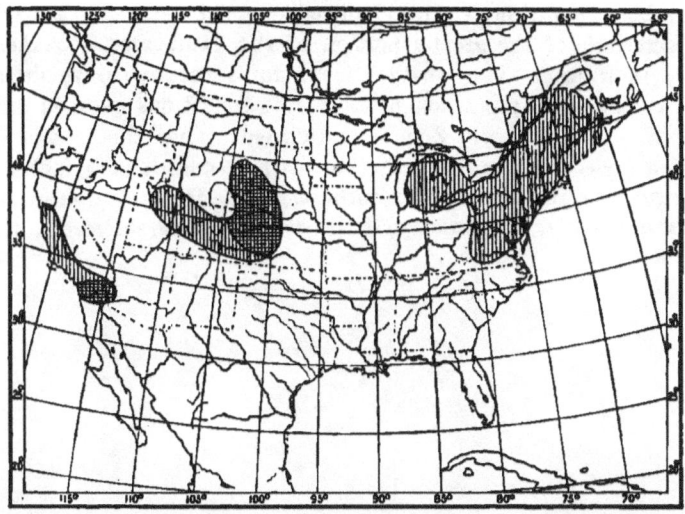

FIG. 44.—Weather. Fifth Day.

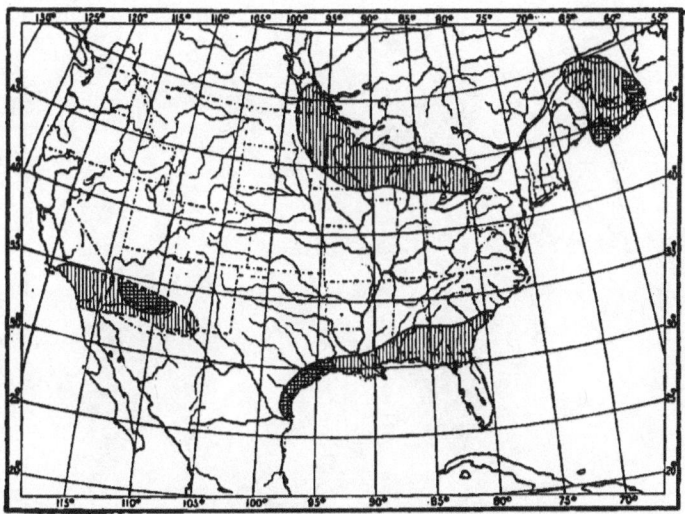

FIG. 45.—Weather. Sixth Day.

another, and of the variability of our weather from day to day in different parts of the United States, are emphasized by these charts of weather conditions. This changeableness of weather is a marked characteristic of the greater portion of the Temperate Zones, especially in winter. The weather maps for successive days do not, as a rule, show a repetition of the same conditions over extended regions. In the Torrid Zone it is different. Over the greater part of that zone the regularity of the weather conditions is such that, day after day, for weeks and months, the same features are repeated. There monotony, here variety, is the dominant characteristic of the weather.

PART IV. — THE CORRELATIONS OF THE WEATHER ELEMENTS AND WEATHER FORECASTING.

CHAPTER IX.

CORRELATION OF THE DIRECTION OF THE WIND AND THE PRESSURE.

THE study of the series of weather maps in Chapters V–VIII has made it clear that some fairly definite relation exists between the general flow of the winds and the distribution of pressure. We now wish to obtain some more definite result as to the relation of the direction of the wind and the pressure. In doing this it is convenient to refer the wind direction to the *barometric* or *pressure gradient* at the station at which the observation is made. The barometric gradient, it will be remembered, is the line along which there is the most rapid change of pressure, and lies at right angles to the isobars (Chapter VII).

Take a small piece of tracing paper, about 3 inches square, and draw upon it a diagram similar to the one here shown. Select the station (between two isobars on any weather map) at which you intend to make your observation. Place the center of the tracing paper diagram over the station, with the dotted line along the

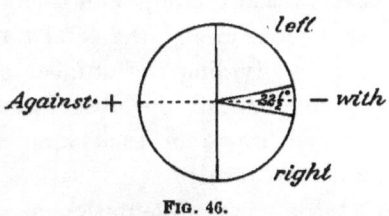

FIG. 46.

barometric gradient, the minus end of the line being towards the area of low pressure. Observe into which of the four sectors (marked *right, left, with, against*) the wind arrow at the station points. Keep a record of the observation. Repeat the observation at least 100 times, using different stations, on the same map or on different maps. Tabulate your results according to the following scheme, noting in the first column the date of the map, in the second, third, fourth, and fifth columns the number of winds found blowing *with*, to the *right* or *left* of, and *against*, the gradient.

TABLE I. — CORRELATION OF THE DIRECTION OF THE WIND AND THE PRESSURE.

Dates	With	Right	Left	Against
Sums				
Percentages				

At the bottom of each column write down the number of cases in that column, and then determine the percentages which these cases are of the total number of observations. This is done by dividing the number of cases in each column by the sum-total of all the observations. When you have obtained the percentage of each kind of wind direction, you have a numerical result.

A graphical presentation of the results may be made by laying off radii corresponding in position to those which divide the sectors in Fig. 46, and whose lengths are proportionate to the percentages of the different wind directions in the table. Thus, for a percentage of 20, the radii may be made 1 inch long, for 40%, 2 inches, etc. When completed, the relative sizes of the sectors will show the relative frequencies of winds blowing in the four different directions with reference to the gradient, as is indicated in Fig. 47.

The Deflection of the Wind from the Gradient: Ferrel's Law. — The law of the deflection of the wind prevailingly to the right of the gradient is known as *Ferrel's Law*, after William Ferrel, a noted American meteorologist, who died in 1891. The operation of this law has already been seen in the spiral circulation of the winds around the cyclone and the anticyclone, as shown on the maps of our series. In the case of the cyclone the gradient is directed inward towards the center; in the case of the anticyclone the gradient is directed outward from the center. In both

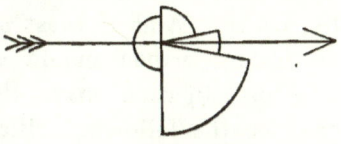

FIG. 47.

cases the right-handed deflection results in a spiral whirl, inward in the cyclone, outward in the anticyclone. The operation of this law is further seen in the case of the *Northeast Trade Winds*. These winds blow from about Lat. 30° N. towards the equator, with wonderful regularity, especially over the oceans. Instead of following the gradient and blowing as north winds, these trades turn to the right of the gradient and become *northeast* winds, whence their name. From about Lat. 30° N. towards the North Pole there is another great flow of winds over the earth's surface. These winds do not flow due north, as south winds. They turn to the right, as do the trades, and become southwest or west-southwest winds, being known as the *Prevailing Westerlies*. Ferrel's Law thus operates in the larger case of the general circulation of the earth's atmosphere, as well as in the smaller case of the local winds on our weather maps.

CHAPTER X.

CORRELATION OF THE VELOCITY OF THE WIND AND THE PRESSURE.

PREPARE a scale of latitude degrees, as explained in Chapter V. Select some station on the weather map at which there is a wind arrow, and at which you wish to study the relation of wind velocity and pressure. Find the rate of pressure change per degree as explained in Chapter VII. Note also the velocity, in miles per hour, of the wind at the station. Repeat the operation 100 or more times, selecting stations in different parts of the United States. It is well, however, to include in one investigation either interior stations alone (*i.e.*, more than 100 miles from the coast) or coast stations alone, as the wind velocities are often considerably affected by proximity to the ocean. And, if coast stations are selected, either onshore or offshore winds should alone be included in one exercise. The investigation may, therefore, be carried out so as to embrace the following different sets of operations:—

A. Interior stations.
B. Coast stations with onshore winds.
C. Coast stations with offshore winds.

Enter your results in a table similar to the one here given:—

TABLE II.— CORRELATION OF WIND VELOCITY AND BAROMETRIC GRADIENT.

For interior (or coast) stations, with onshore (or offshore) winds, in the United States during the month (or months) of

Rates of Pressure Change per Latitude Degree	∞–20	20–10	10–5	5–3½	3½–2½	2½–2	etc.
Distances between Isobars in Latitude Degrees	0–½	½–1	1–2	2–3	3–4	4–5	etc.
Wind Velocities (miles per hour)							
Sums							
Cases							
Means							

The wind velocity for each station is to be entered in the column at whose top is the rate of pressure change found for that station. Thus, if for any station the rate of pressure change is $3\frac{1}{3}$ (*i.e.*, .03 inch in one latitude degree), and the wind velocity at that station is 17 miles an hour, enter the 17 in the fourth and fifth columns of the table. When you find that the rate of pressure change for any station falls into two columns of the table, as, *e.g.*, 10, or 5, or $3\frac{1}{3}$, then enter the corresponding wind velocity in both those columns.

In the space marked *Sums* write the sum-total of all the wind velocities in each column. The *Cases* are the number of separate observations you have in each column. The *Means* denote the average or mean wind velocities found in each column, and are obtained by dividing the sums by the cases.

Study the results of your table carefully. Deduce from your own results a general rule for wind velocities as related to barometric gradients.

The dependence of wind velocities on the pressure gradient is a fact of great importance in meteorology. The ship captain at sea knows that a rapid fall of his barometer means a rapid rate of pressure change, and foretells high winds. He therefore makes his preparations accordingly, by shortening sail and by making everything fast. The isobaric charts of the globe for January and July show that the pressure gradients are stronger (*i.e.*, the rate of pressure change is more rapid) over the Northern Hemisphere in January than in July. This fact would lead us to expect that the velocities of the general winds over the Northern Hemisphere should be higher in winter than in summer, and so they are. Observations of the movements of clouds made at Blue Hill Observatory, Hyde Park, Mass., show that the whole atmosphere, up to the highest cloud level, moves almost twice as fast in winter as in summer. In the higher latitudes of the Southern Hemisphere, where the barometric gradients are prevailingly much stronger than in the Northern, the wind velocities are also prevailingly higher than they are north of the equator. The prevailing westerly winds of the Southern Hemisphere, south of latitude of 30° S., blow with high velocities nearly all the time, especially during the winter months (June, July, August). These winds are so strong from the westward that vessels trying to round Cape Horn from the east often occupy weeks beating against head gales, which continually blow them back on their course.

CHAPTER XI.

FORM AND DIMENSIONS OF CYCLONES AND ANTICYCLONES.

A. **Cyclones.**—Provide yourself with a sheet of tracing paper about half as large as the daily weather map. Draw a straight line across the middle of it; mark a dot at the center of the line, the letter N at one end, and the letter S at the other. Place the tracing paper over a weather map on which there is a fairly well enclosed center of low pressure (*low*), having the dot at the center of the *low*, and the line parallel to the nearest meridian, the end marked N being towards the top of the map. When thus placed, the paper is said to be *oriented*. Trace off the isobars which are nearest the center. In most cases the 29.80-inch isobar furnishes a good limit, out to which the isobars may be traced. Continue this process, using different weather maps, until the lines on the tracing paper begin to become too confused for fairly easy seeing. Probably 15 or 20 separate areas of low pressure may be traced on to the paper. It is important to have all parts of the cyclonic areas represented on your tracing. If most of the isobars you have traced are on the southern side of cyclones central over the Lakes or lower St. Lawrence, so that the isobars on the northern sides are incomplete, select for your further tracings weather maps on which the cyclonic centers are in the central or southern portions of the United States, and therefore have their northern isobars fully drawn.

When your tracing is finished you have a *composite portrait* of the isobars around several areas of low pressure. Now study the results carefully. Draw a heavy pencil or an ink line on the tracing paper, in such a way as to enclose the average area outlined by the isobars. This average area will naturally be of smaller dimensions than the outer isobars on the tracing paper, and of larger dimensions than the inner isobars, and its form will follow the general trend indicated by the majority of the isobars, without reproducing any exceptional shapes.

Write out a careful description of the average *form, dimensions* [measured by a scale of miles or of latitude degrees (70 miles = 1 degree about)] and *gradients* of these areas of low pressure, noting any tendency to elongate in a particular direction; any portions of the composite where the gradients are especially strong, weak, etc.

B. **Anticyclones.** — This investigation is carried out in precisely the same manner as the preceding one, except that anticyclones (*highs*) are now studied instead of cyclones. The isobars may be traced off as far away from the center as the 30.20-inch line in most cases. When, however, the pressure at the center is exceptionally high, it will not be necessary to trace off lower isobars than those for 30.30, or 30.40, or sometimes 30.50 inches.

Loomis's Results as to Form and Dimensions of Cyclones and Anticyclones. — One of the leading American meteorologists, Loomis, who was for many years a professor in Yale University, made an extended study of the form and dimensions of areas of low and high pressure as they appear on our daily weather maps. In the cases of areas of low pressure which he examined, the average form of the areas was elliptical, the longer diameter being nearly twice as long as the shorter (to be exact the ratio was 1.94:1). The average direction of the longer diameter he found to be about NE. (N. 36° E.), and the length of the longer diameter often 1600 miles. In the case of areas of high pressure, Loomis also found an elliptical form predominating; the longer diameter being about twice as long as the shorter (ratio 1.91:1), and the direction of trend about NE. (N. 44° E.). These characteristics hold, in general, for the cyclonic

98 WEATHER ELEMENTS AND FORECASTING.

and anticyclonic areas of Europe also. The cyclones of the tropics differ considerably from those of temperate latitudes in being nearly circular in form.

CHAPTER XII.

CORRELATION OF CYCLONES AND ANTICYCLONES WITH THEIR WIND CIRCULATION.

A. **Cyclones.** — Something as to the control of pressure over the circulation of the wind has been seen in the preliminary exercises on the daily weather maps. We now proceed to investigate this correlation further by means of the composite portrait method. This method is a device to bring out more clearly the general systems of the winds by throwing together on to one sheet a large number of wind arrows in their proper position with reference to the controlling center of low pressure. In this way we have many more observations to help us in our investigation than if we used only those which are given on one weather map, and the circulation can be much more clearly made out.

Provide yourself with a sheet of tracing paper, prepared as described in Chapter XI. Place the paper over an area of low pressure on some weather map, with the dot at the center of the *low*, and having the paper properly oriented, as already explained. Trace off all the wind arrows around the center of low pressure, making the lengths of these arrows roughly proportionate (by eye) to the velocity of the wind, according to some scale previously determined upon. Include on your tracing all the wind arrows reported at stations whose lines of pressure-decrease converge towards the low pressure center. Repeat this operation, using other centers of low pressure on other maps, until the number of arrows on the tracing paper is so great that the composite begins to become confused. Be careful always to center and orient your tracing paper properly.

Select the weather maps from which you take your wind arrows so that the composite shall properly represent winds in all parts of the cyclonic area.

Deduce a general rule for the circulation and velocity of the wind in a cyclonic area, as shown on your tracing, and write it out.

B. **Anticyclones.** — This exercise is done in precisely the same way as the preceding one, except that anticyclones and their winds are studied instead of cyclones.

Deduce a general rule for the circulation and velocity of the wind in an anticyclonic area, as shown on your tracing, and write it out.

The control of the wind circulation by areas of low and high pressure is one of the most important laws in meteorology. Buys-Ballot, a Dutch meteorologist, first called attention to the importance of this law in Europe, and it has ever since been known by his name. Buys-Ballot's Law is generally stated as follows: *Stand with your back to the wind, and the barometer will be lower on your left hand than on your right*.[1] This statement, as will be seen, covers both cyclonic and anticyclonic systems. The circulations shown on your tracings hold everywhere in the Northern Hemisphere, not only around the areas of low and high pressure seen on the United States weather maps, but around those which are found in Europe and Asia, and over the oceans as well. Mention has already been made, in the chapter on isobars (VII), of the occurrence of immense cyclonic and anticyclonic areas, covering the greater portion of a continent or an ocean, and lasting for months at a time. These great cyclones and anticyclones have the same systems of winds around them that the smaller areas, with similar characteristics, have on our weather maps. A further extension of what has just been learned will show that if in any region there comes a change from low pressure to high pressure, or *vice versa*, the system of winds in that region will also change. Many such changes of pressures and winds actually occur in different parts of the world, and are of great importance in controlling the climate. The best-known and the most-marked of all these changes occurs in the case of

---

[1] In the Northern Hemisphere.

India. During the winter, an anticyclonic area of high pressure is central over the continent of Asia. The winds blow out from it on all sides, thus causing general northeasterly winds over the greater portion of India. These winds are prevailingly dry and clear, and the weather during the time they blow is fine. India then has its dry season. As the summer comes on, the pressure over Asia changes, becoming low; a cyclonic area replaces the winter anticyclone, and inflowing winds take the place of the outflowing ones of the winter. The summer winds cross India from a general southwesterly direction, come from over the ocean, and are moist and rainy. India then has its rainy season. These seasonal winds are known as *Monsoons*, a name derived from the Arabic and meaning *seasonal*.

The accompanying figure (Fig. 48) is taken from the *Pilot Chart of the North Atlantic Ocean*, published by the Hydrographic Office of the United States Navy for the use of seamen. It shows the wind circulation around the center of a cyclone which is moving northward along the Atlantic Coast of the United States. The long arrow indicates the path of movement; the shorter arrows indicate the directions of the winds. By means of such a diagram as this a captain is able to calculate, with a considerable degree of accuracy, the position of the center of the cyclone, and can often avoid the violent winds near that center by sailing away from it, or by "lying to," as it is called, and waiting until the center passes by him at a safe distance. These cyclones which come up the eastern coast of the United States at certain seasons are usually violent, and often do considerable damage to shipping. The Weather Bureau gives all the warning possible of the coming of these *hurricanes*, as they are called, by displaying *hurricane signals* along the coast, and by issuing telegraphic warnings to newspapers. In this way ship captains, knowing of the approach of gales dangerous to navigation, may keep their vessels in port until all danger is past. Millions of dollars' worth of property and hundreds of lives have thus been saved.

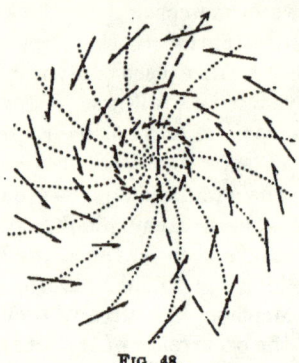

FIG. 48.

CHAPTER XIII.

CORRELATION OF THE DIRECTION OF THE WIND AND THE TEMPERATURE.

It is evident, from even the most general observation of the weather elements, that the temperature experienced at any place is very largely dependent upon the direction of the wind. Thus, for instance, in the United States, a wind from some northerly point is likely to bring a lower temperature than a southerly wind. To investigate this matter more closely, and to discover how the winds at any station during any month are related to the temperatures noted at that station, we proceed as follows: —

Select the Weather Bureau station at which you wish to study these conditions. Note the direction of the wind and the temperature at that station on the first day of any month. Prepare a table similar to the following one.

TABLE III. — CORRELATION OF THE DIRECTION OF THE WIND AND THE TEMPERATURE.

At .... during the Month of

WIND DIRECTIONS	N.	NE.	E.	SE.	S.	SW.	W.	NW.	
TEMPERA-TURES									
Sums									Total
Cases									Total
Means									Mean

Enter the temperature at 8 A.M. on the first day of the month in a column of the table under the proper wind direction. Thus, if the wind is NE., and the temperature 42°, enter 42 in the second column of the table. Repeat the observation for the same station, and for all the other days of the month, recording the temperatures in each case in their appropriate columns in the table. Omit all cases in which the wind is *light*, because winds of low velocities are apt to be considerably affected by local influences. When the observations for the whole month have been entered in the table, add up all the temperatures in each column (*sums*). Find the mean temperature (*means*) observed with each wind direction by dividing the sums by the number of observations in each column (*cases*). Add all the sums together; divide by the total number of cases, and the result will be the mean temperature* for the month at the station. The general effect of the different wind directions upon the temperature is shown by a comparison of the means derived from each column with the mean for the month.

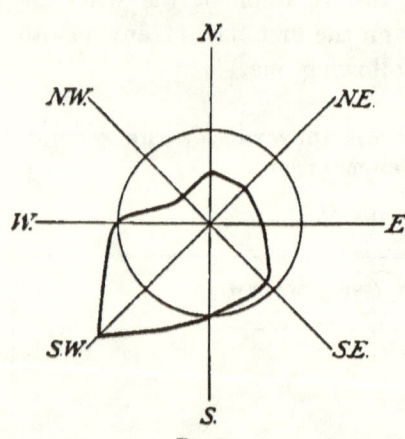

Fig. 49.

A graphic representation of the results of this investigation will help to emphasize the lesson. Draw, as in the accompanying figure (Fig. 49), eight lines from a central point, each line to represent one of the eight wind directions. About the central point describe a circle, the length of whose radius shall correspond to the mean temperature of the month, measured on some convenient scale. Thus, if the mean

---

* Derived from the 8 A.M. observations. This does not give the true mean temperature.

temperature of the month is 55° and a scale of half an inch is taken to correspond to 10° of temperature, the radius of the circle must be five and a half times half an inch, or 2¾ inches. Next lay off on the eight wind lines the mean temperatures corresponding to the eight different wind directions, using the same scale (½ in. = 10°) as in the previous case. Join the points thus laid off by a heavy line, as shown in Fig. 49. The figure, when completed, gives at a glance a general idea of the control exercised by the winds over the temperatures at the station selected. Where the heavy line crosses a wind line *inside* the circle it shows that the average temperature accompanying the corresponding wind direction is below the mean. When the heavy line crosses any wind line *outside* the circle, it shows that the average temperature accompanying the corresponding wind direction is above the mean. Such a figure is known as a *wind rose*.

The cold wave and the sirocco are two winds which exercise marked controls over the temperature at stations in the central and eastern United States. The *cold wave* has already been described in Chapter V. It is a characteristic feature of our winter weather. It blows down from our Northwestern States or from the Canadian Northwest, on the western side of a cyclone. It usually causes sudden and marked falls in temperature, sometimes amounting to as much as 50° in 24 hours. The *sirocco* is a southerly or southwesterly wind. It also blows into a cyclone, but on its southern or southeastern side. Coming from warmer latitudes, and from over warm ocean waters, the sirocco is usually a warm wind, in marked contrast to the cold wave. In winter, in the Mississippi Valley and on the Atlantic Coast, the sirocco is usually accompanied by warm, damp, cloudy, and snowy or rainy weather. The high temperatures accompanying it (they may be as high as 50° or 60° even in midwinter) are very disagreeable. Our warm houses and our winter clothing become oppressive and we long for the bright, crisp, cold weather brought by the *cold wave*. In summer when a sirocco blows we have our hottest spells. Then sunstrokes and prostrations by the heat are most common, and our highest temperatures are recorded. The word *sirocco* (from *Syriacus* = Syrian) was first used as the name of a warm southerly wind in Italy. The cause and the characteristics of the Italian sirocco and of the American sirocco are similar, and the name may therefore be applied to our wind as well as to the Italian one. In the Southern Hemisphere, at Buenos Ayres, Argentine Republic, there is a similar contrast between two different winds. The *pampero* is similar in many respects to our cold wave. It is a dry, cool, and refreshing wind, blowing over the vast level stretches of the Argentine pampas from the southwest. The *norte* is a warm, damp, depressing northerly wind corresponding to our sirocco.

CHAPTER XIV.

CORRELATION OF CYCLONES AND ANTICYCLONES AND THEIR TEMPERATURES.

A. **Cyclones.** — It follows from the two preceding exercises that some fairly definite distribution of temperature, depending upon the wind direction, should exist around areas of low and high pressure. Try to predict, on the basis of the results obtained in Chapters XII and XIII, what this relation of temperatures and cyclones and anticyclones is. Then work out the relation independently of your prediction, by studying actual cases obtained from the weather maps, as follows: —

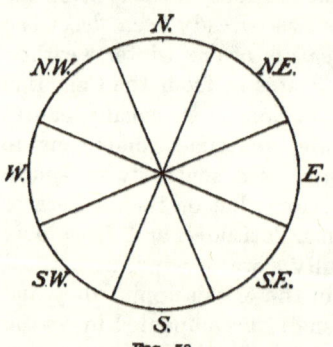

FIG. 50.

Prepare a sheet of tracing paper as shown in Fig. 50. The diameter of the circle should be sufficiently large to include within the circle the average area covered by a cyclone on the weather maps. Place the tracing paper, properly divided in accordance with the figure, over a well-defined area of low pressure on a weather map, centering and orienting it carefully.

Take the temperature at the station which lies nearest the center of the figure as the standard. Notice the temperatures at all the other stations which fall within the limits of the circle, and mark down at the proper places on the tracing paper, the + or − departures of these local temperatures from the standard temperature. Thus, if the standard is 37°, and a station has a temperature of 46°, enter + 9° at the proper place on your tracing paper. Again, if a certain station has 24°, enter − 13° at the proper place on the paper. Continue this process until your paper has all of its divisions well filled. It is best to select all the maps used in this investigation from the same month, for in that case the data are more comparable than if different months are taken. When a sufficient number of examples has been obtained, find the average departure(+ or −) of the temperatures in each division of the tracing from the central standard temperature. Express these averages graphically by means of a *wind rose*, as in the last exercise.

Another Method. — The above correlation may be investigated by means of another method, as follows : —

Prepare a piece of tracing paper by drawing an N. and S. line upon it, and placing a dot at the center of the line. Lay the paper over an area of low pressure on any weather map, centering and orienting it properly, as in the previous exercises. Trace off the isotherms which are near the center of low pressure. Repeat this process with several maps, selecting different ones from those used in the first part of this exercise. Formulate a rule for the observed distribution of temperature, and determine the reasons for this distribution. Note carefully any effects of the cyclone upon the temperature gradient.

B. **Anticyclones.** — The correlation of anticyclones with their temperatures is studied in precisely the same way as the preceding correlation. Both methods suggested in the case of cyclones should be used in the case of anticyclones. When your results have been obtained, formulate a general rule for the observed distribution of temperature in anticyclones, and determine the reasons for this distribution.

Find from your composites the average temperature of cyclones and of anticyclones, and compare these averages.

The unsymmetrical distribution of temperature around cyclones, which is made clear by the foregoing exercises, is very characteristic of these storms in our latitudes, and especially in the eastern United States. That this has an important effect upon weather changes is evident, and will be further noted in the chapter on *Weather Forecasting*. The cyclones which begin over the oceans near the equator at certain seasons, and thence travel to higher latitudes, — *tropical cyclones*, so called, — differ markedly from our cyclones in respect to the distribution of temperature around them. The temperatures on all sides of tropical cyclones are usually remarkably uniform, the isotherms coinciding fairly closely with the isobars. The reason for this is to be found in the remarkable uniformity of the temperature and humidity conditions over the surrounding ocean surface, from which the inflowing winds come. In the case of our own cyclones, in the eastern United States, the warm southerly wind, or sirocco, in front of the center has very different characteristics from those of the cold northwesterly wind, or *cold wave*, in the rear, as has become evident through the preceding exercise. These winds, therefore, naturally show their effects in the distribution of the temperatures in different parts of the cyclonic area.

CHAPTER XV.

CORRELATION OF THE DIRECTION OF THE WIND AND THE WEATHER.

SELECT a file of daily weather maps for some month. Commencing with the first map in the set, observe the weather and the direction of the wind at a considerable number of stations in the same general region (as, *e.g.*, the Lake region, the lower Mississippi Valley, the Pacific Coast, etc.). Enter each case in a table, similar to Table IV below, by making a check in the column under the appropriate wind direction and on a line with the appropriate type of weather.

TABLE IV. — CORRELATION OF THE DIRECTION OF THE WIND AND THE WEATHER.

At .... during the Month of

	N.	NE.	E.	SE.	S.	SW.	W.	NW.	TOTALS	PER-CENTAGES
Clear									D	J
Fair									F	K
Cloudy									G	L
Rain and Snow									H	M
Totals									A	
Percentages	B	C	etc.							

Count every observation of *rain* or *snow also as cloudy*, for it usually rains or snows only when the sky is cloudy. Continue your observations on all the maps for the month you have chosen. Then count up the whole number of cases of *clear* weather you have found with north winds, and write down this sum in the first space, in the column reserved for N. winds. Do the same with *fair* and *cloudy* weather. Add up and enter at the bottom of the column in the space marked *Totals* the whole number of observations of *clear, fair*, and *cloudy* weather you have observed with N. winds. Then find what percentage of the weather with N. winds was *clear*, and enter this percentage next to the sum of *clear* weather observations, in the first division in the column headed N. Do the same for *fair, cloudy*, and *rainy* or *snowy* weather, deriving the percentages of rain or snow from the total of *clear* and *fair* and *cloudy*. Repeat this process of summarizing in every column. Your results will then show the percentages of the different kinds of weather noted with the different wind directions.

The lower division of the table and the last two columns on the right are to be used for a general summary of the whole investigation. By adding together all the totals of *clear, fair,* and *cloudy* weather observed with all the eight wind directions you obtain the whole number of observations you have made. Enter this in the space marked A, at the right of the table. From this grand total and the total number of observations in each column you may find (in percentages) the relative frequencies of the different wind directions. These should be entered under the totals at the bottom of each column, in the spaces marked *Percentages* (spaces B, C, etc.). The total number of observations of *clear, fair, cloud,* and *rain* or *snow*, noted with all the wind directions, are to be entered in spaces D, F, G, and H, at the right of the table. From these totals, and from the grand total in space A, we can determine the relative frequency (in percentages) of each kind of weather during the month. These results should be entered in spaces J, K, L, and M.

Study these results carefully. Formulate them in a brief written statement. Express graphically the following: —

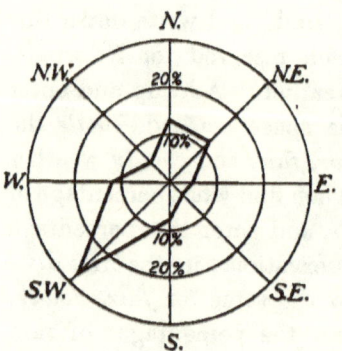

FIG. 51.

A. The percentages of frequency of the different wind directions during the month.

B. The percentages of the different kinds of weather noted during the whole month for all wind directions.

A wind rose, indicating the percentages of frequency of different winds during a month, or

CYCLONES AND ANTICYCLONES. 109

a year, or several years, may be constructed as shown in Fig. 51.

A certain convenient scale is adopted as representing a frequency of 10%, and a circle is drawn with this unit as a radius. A second circle, with a radius twice as long, represents a frequency of 20%, and a third circle, with a radius three times as long, represents a frequency of 30%. Additional circles may be added if necessary. Distances corresponding to the different percentages of frequency of the eight wind directions are then laid off along the eight radii of the circles, and the points thus fixed are joined by a line.

The results asked for under question *B* may be plotted as a weather rose on a diagram similar to that above figured. In this case the percentages of frequency of the different varieties of weather (*clear, fair, cloudy, stormy*) may be indicated on the same figure by using different kinds of lines. Thus, a *solid* line may be employed to represent *clear* weather; a *broken* line for *fair;* a *broken and dotted* line for *cloudy;* and a *dotted* line for *stormy* weather.

CHAPTER XVI.

CORRELATION OF CYCLONES AND ANTICYCLONES AND THE WEATHER.

A. **Cyclones.** — Prepare a piece of tracing paper as shown in Fig. 52, making the diameter of the outer circle about 1000 miles[1] and of the inner circle 500 miles. Place this diagram over a cyclone on any weather map, centering and orienting it carefully. Trace off the weather signs (indicating *clear, fair, cloudy, rain* or *snow*) around the cyclonic center from the map on to the tracing paper, taking only observations which are not more than halfway from the cyclonic center to the neighbor-

---

[1] Use the scale of miles given on the weather map.

ing anticyclonic center. Repeat this process with successive weather maps until the diagram is well filled in all its different divisions.

A. Draw a line on the tracing paper enclosing the average area of *cloud* (including *rain and snow*), and a second line enclosing the average area of *precipitation* (*rain* or *snow*).

B. Determine the percentages of *clear, fair, cloudy*, and stormy observations for each division of the tracing paper, *i.e.*, (*a*) for the eight sectors of the large circle; (*b*) for the whole of the small circle; and (*c*) for the portion of the diagram between the circumference of the inner circle and the circumference of the outer circle.

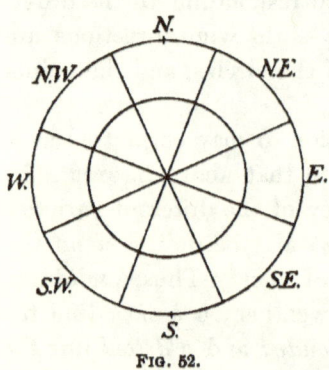

Fig. 52.

C. Write out in general terms a description of the weather distribution in cyclones as illustrated by your own investigation.

B. **Anticyclones.** — This exercise is done in the same way as the preceding one, except that anticyclones are substituted for cyclones.

The association of fair weather with anticyclones and of stormy weather with cyclones is one of the most important lessons learned from a study of the weather maps. The great areas of high and low pressure control our weather. On land, where daily weather maps are so easily accessible, a glance at the map serves in most cases to give a fairly accurate idea of the position and extent of cyclones and anticyclones, and hence also of the distribution of weather. At sea, on the other hand, the navigator has no daily weather maps to refer to, and his knowledge of the weather conditions which he may expect must be gained from his own observations alone. Of these local observations, the pressure readings are by far the most important. A falling barometer usually means the approach of a cyclone, with wind, or storm, or both. A rising barometer, on the other hand, is usually an indication of the fine weather associated with an anticyclone. The unsymmetrical distribution of weather, characteristic of our cyclones in the United States, and also of most cyclones in the Temperate Zone, is associated with their unsymmetrical form, and the unsymmetrical distribution of their temperature already studied. Tropical cyclones have a wonderfully uniform distribution of weather on all sides of their centers, just as they have a symmetrical form and an even temperature distribution all around them.

CHAPTER XVII.

PROGRESSION OF CYCLONES AND ANTICYCLONES.

So far no definite study has been made of the changes in the positions of cyclones and anticyclones. If these areas of stormy and fair weather always occupied the same geographic positions, the different portions of the country would always have the same kinds of weather. A knowledge of the movements of the areas of low and high pressure makes weather forecasting possible.

A. **Cyclones.** — Select a set of daily weather maps for a month. Turn to the first map of the series. Note the position of the center of low pressure, and indicate this position on a blank weather map of the United States by marking down a small circle at the proper place. If there are two or more areas of low pressure on the map, indicate the position of each one of them in a similar way. Turn to the second map of the series, and again enter on the blank map the position of the center of low pressure. Connect the two positions of each center by a line. This line may be called the *track* of the low pressure center. Continue this process through the whole set of maps, connecting all the new positions with the last positions of their respective centers. Mark each position with the appropriate date in small, neat figures. When completed your map will show at a glance the tracks followed by all the cyclones which traveled across the United States during the month you selected. Study these tracks carefully. Notice whether there is any prevailing direction in which the cyclones move, and whether they show any preference for particular paths across the country. Can you frame a general rule for the prevailing direction and path of movement? Are there any cases which do not accord with the rule? If so, describe them. In what position, with reference to the cyclonic tracks during the month you are studying, is the region in which you are now living?

Next determine the *velocities* with which these cyclones moved. Prepare a scale of latitude degrees, as described in Chapter V, or of miles, as given at the bottom of the weather map. Measure the distances, in miles, between the successive positions of all the cyclonic centers. Divide these distances by 24 in order to obtain the velocity in *miles per hour*. What is the highest velocity per hour with which any cyclone moved during the month? What is the lowest? What is the mean, or average, velocity?

Study the tracks and velocities of cyclones in a similar way during several other months. Compare the positions of the tracks, and the velocities of progression, in summer and winter.

B. **Anticyclones.** — Study the tracks and velocities of anticyclones in precisely the same manner. Compare the results derived from your investigations in the two cases.

Cyclonic Tracks and the Prevailing Westerly Winds. — The correspondence in the direction of movement of most of our cyclones and of the prevailing westerly winds of the Northern Hemisphere (see Chapter IX) will readily be noted. Our weather maps show us the atmospheric conditions over the United States alone, and we can therefore trace the progression of our cyclones over but a limited area of the Northern Hemisphere. An examination of the daily weather maps of the North Atlantic and North Pacific Oceans, which are based on observations made on board ships, and of the weather maps of other countries, shows us that these atmospheric disturbances which appear on our maps may often be traced for long distances across oceans and lands, and that they in reality form a great procession across the northern portion of this hemisphere, and towards and around the North Pole.

The average velocity of cyclones in the United States was carefully determined by Loomis. His observations show that the mean hourly velocity of cyclones for the entire year is 28.4 miles, the maximum (34.2 miles) coming in February, and the minimum (22.6 miles) in August. Over the North Atlantic Ocean the hourly velocity is 18 miles; in Europe, 16.7 miles. The greater velocity of cyclonic movement in the United States in winter recalls what was said at the close of Chapter X concerning the steeper barometric gradient and the more rapid movement of the whole atmosphere over the Northern Hemisphere in winter.

CHAPTER XVIII.

SEQUENCE OF LOCAL WEATHER CHANGES.

THE next, and last, step in our study of the correlation of the various weather elements concerns the sequence of weather changes at a station before, during, and after the passage of a cyclone and of an anticyclone.

A. **Cyclones.** — I. Select some station which the weather maps show to have been directly on the track of a well-developed cyclone, *i.e.*, to have been passed over by the center of the cyclone. Note the weather conditions at this station, before, during, and after the passage of the storm. Tabulate your observations according to the following scheme: —

TABLE V.

Weather Changes at .... during ....

DATES	PRESSURE	TEMPERATURE	WIND DIRECTION	WIND VELOCITY	WEATHER	DIREC. AND DIST. OF STORM CENTER

In the last column of the table enter the direction and the distance of the cyclonic center from the station, at each observation.

II. Select a station which was north of the track of a cyclone, and tabulate (in a separate table) the weather conditions at that station before, during, and after the passage of the center.

III. Do the same for a station which was south of the track of a cyclone. Repeat these observations for several stations.

B. **Anticyclones.** — Make a similar series of observations for the passage of an anticyclone centrally over, north and south of, several stations.

Study the sequence of the weather changes shown in the various tables. Deduce a general rule for these changes and write it out.

CHAPTER XIX.

WEATHER FORECASTING.

In a letter dated at Philadelphia, July 16, 1747, Benjamin Franklin wrote to his friend Jared Eliot as follows: "We have frequently along the North American coast storms from the northeast which blow violently sometimes three or four days. Of these I have had a very singular opinion for some years, viz.: that, though the course of the wind is from northeast to southwest, yet the course of the storm is from southwest to northeast; the air is in violent motion in Virginia before it moves in Connecticut, and in Connecticut before it moves at Cape Sable, etc. My reason for this opinion (if the like have not occurred to you) I will give in my next."

In a second letter to the same correspondent, dated Philadelphia, Feb. 13, 1749-50, Franklin states his reasons as follows: "You desire to know my thoughts about the northeast storms beginning to leeward. Some years since, there was an eclipse of the moon at nine o'clock in the evening, which I intended to observe; but before night a storm blew up at northeast, and continued violent all night and all the next day; the sky thick-clouded, dark, and rainy, so that neither moon nor stars could be seen. The storm did great damage all along the coast, for we had accounts of it in the newspapers from Boston, Newport, New York, Maryland, and Virginia; but what surprised us was to find in the Boston newspapers an account of the observation of that eclipse made there; for I thought as the storm came from the northeast it must have begun sooner at Boston than with us, and consequently have prevented such an observation. I wrote to my brother about it, and he informed me that the eclipse was over there an hour before the storm began. Since which time I have made inquiries from time to time of travelers, and observed the accounts in the newspapers from New England, New York, Maryland, Virginia, and South Carolina; and I find it to be a constant fact that northeast storms begin to leeward, and are often more violent there than to windward" (Sparks's *Life of Franklin*, VI, 79, 105, 106).

The fact that our northeast storms come from the southwest, which was first noticed by Benjamin Franklin some years before he put the suggestion just quoted in writing, was one of the great contributions to meteorology made by Americans. Modern weather forecasting essentially depends upon the general eastward movement of cyclones and anticyclones, with their accompanying weather conditions.

The daily weather map shows us the actual condition of the weather all over the United States at 8 A.M., " Eastern Standard Time." The positions of cyclones and of anticyclones; of areas of clear, fair, cloudy or stormy weather, and of regions of high or low temperatures, are plainly seen at a glance. These areas of fair and foul weather, with their accompanying systems of spiralling winds, move across country in a general easterly direction. Knowing something of their direction and rate of movement, we can determine, with greater or less accuracy, their probable positions in 12, 24, 36, or 48 hours. The prediction or foretelling of the weather which may be expected to prevail at any station or in any district is *weather forecasting*.

Weather forecasts are usually made on our daily weather maps for 24 hours in advance. It is by no means an easy thing to make accurate weather forecasts. Careful study and much practice are required of the forecasters of the Weather Bureau before they are permitted to make the official forecasts which are printed on the daily maps and in the newspapers.

A simple extension and application of the principles learned through the preceding exercises make it possible for us to forecast coming weather changes in a general way. These suggestions are, however, not at all to be considered as a complete discussion of this complicated problem.

Weather forecasts include the probable changes in *temperature*, *wind direction* and *velocity*, and *weather*. Pressure is not included. Begin your practice in weather forecasting by considering only the changes that may be expected at your own point of observation, and at first confine yourself to predicting temperature changes alone.

Temperature. — Provide yourself with a blank weather map. Draw an isotherm east and west across the map, through your station. Draw a few other isotherms all the way across the map, parallel with the first one, and so arranged that they will be equal distances apart, the most northerly one running through northern Maine and the Northwestern States, and the most southerly one through southern Florida and Texas. Recalling what you have already discovered concerning the eastward movement of our weather conditions, what forecast will you make as to the coming temperatures at your station? Add some additional east and west isotherms, so that there will be twice as many on your map as before. What effect will this change in the temperature distribution on your map have upon the temperature forecast you make for your station? Formulate a general rule as to temperature forecasts under the conditions of isothermal arrangement here suggested.

On a second blank weather map draw an isotherm through your station inclined from northwest to southeast. Draw a few other isotherms parallel to the first, and each one representing a temperature 10° higher than that indicated by the adjacent isotherm on the east. Make a general forecast of the temperature conditions that may be expected at your station, as to *kind of change*, if any; *amount of change*, and *rapidity of change*. Of the isotherms just drawn, erase every second one; still, however, letting those that are left represent differences of temperature of 10°. What forecast will you now make as to temperature? How does this forecast compare with that just made?

Now draw twice as many isotherms on your map as you had in the first place, still letting these lines represent differences of temperature of 10° in each case. Make a forecast of the kind, amount, and rapidity of temperature change at your station under the conditions represented on this map. How does this forecast compare with the two just made? Formulate a general rule governing temperature forecasts in cases of isothermal arrangement such as those here considered.

Take another blank map. Draw through your station an isotherm inclined from northeast to southwest. Draw other isotherms parallel to this, west of your station, letting each successive isotherm represent a temperature 10° lower than that indicated by the adjacent isotherm on the east. Make a temperature forecast for your station under these conditions. Diminish and increase the number of isotherms on your map, as suggested in the preceding example, making temperature forecasts in each case, and comparing the three sets of forecasts. Formulate a general rule for temperature forecasts made under these systems of isotherms.

Make temperature forecasts from the daily weather maps for your own station, using the knowledge that you have already gained as to the progression of cyclones and anticyclones (Chapter XVII), and as to the temperature distribution in these areas (Chapter XIV), to help you in this work. Study each day's map carefully before you decide on what you will say. Then write out your own forecast, and afterwards compare your forecast with that made by the Weather Bureau. Note also, by reference to your own instrumental observations, whether the succeeding temperature conditions are such as you predicted.

Wind Direction.—The weather maps already studied taught us that our winds habitually move in spirals. The composite picture of the wind circulation around cyclones and anticyclones (Chapter XII) further emphasized this important fact. Evidently this law of the systematic circulation of the winds around centers of low and high pressure may be utilized in making forecasts of wind direction.

Applying the knowledge already gained concerning cyclonic and anticyclonic wind circulations, ask yourself what winds a station should have which is within the range of the cyclonic wind system, and is in the following positions with reference to the center: *northeast, north, northwest, east, at the center, west, southeast, south, southwest*. Ask yourself precisely the same questions with reference to a station within an anticyclonic wind system. Write out a general rule for the kinds of wind changes which may be expected to take place under these different conditions.

When a station is south of the track of a passing cyclone its winds are said to *veer*, and the change in the direction of its winds is called *veering*. A station north of the track of a passing cyclone has a change of direction in its winds which is known as *backing*, the winds themselves being said to *back*.

Wind Velocity.—What general relation between wind velocities and areas of low and high pressure did you discover in your study of the weather maps? What was the result of your work on the correlation of the velocity of the wind and the barometric gradient in Chapter X? And what general statement as to the relation between the velocity of the wind and its distance from a cyclonic or anticyclonic center may be made as the result of your work in Chapter XII, on the correlation of cyclones and anticyclones with their wind circulation? These results must be borne in mind in making predictions of coming changes in wind velocities. Forecasts of wind velocities are made in general terms only, — *light*, *moderate*, *fresh*, *brisk*, *high*, *gale*, *hurricane*, — and are not given in miles per hour.

Make forecasts of wind direction and velocity from the daily weather maps for your own station. Continue these for a week or two, keeping record of the verification or non-verification of each of your forecasts. Then make daily forecasts of temperature and of wind direction and velocity together. Write out your own forecast for each day before you compare it with the official forecast, and if the two differ, keep note of which one seemed to you to be the most accurate.

Weather. — What general relation between kind of weather and cyclones and anticyclones was illustrated on the six maps of our series? What is the average distribution of the different kinds of weather around cyclones and anticyclones, as shown by your composites? (Chapter XVI.) What changes in weather will ordinarily be experienced at a station as a cyclone approaches, passes over, and moves off? What conditions will prevail in an anticyclone?

Make a series of daily forecasts for your own station of probable weather changes, omitting temperature and winds at first. Include in your weather forecasts the state of the sky (*clear*, *fair*, *cloudy*); the changes in the state of the sky (increasing or decreasing cloudiness); the kind of precipitation (*rain* or *snow*) and the amount of precipitation (*light* or *heavy*).

Write out your forecasts; compare them with the official forecasts, and notice how fully they are verified. Then add temperature and winds to your forecasts so that you will make a complete prediction of probable changes in temperature (kind, amount, and rapidity of change), wind (direction and velocity), and weather. Practice making these complete forecasts for several weeks, if time allows. Use all the knowledge that

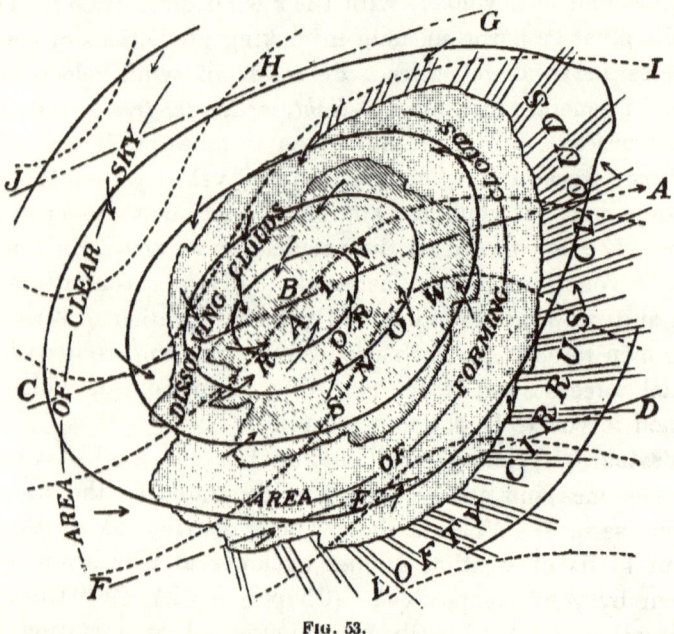

FIG. 53.

you have gained in the preceding work to aid you in this. Study each weather map very carefully. Do not write down your forecast until you are sure that you have done the best you can.

Vary this exercise by extending your forecasts so as to embrace the whole section of country in which your station is situated (as, *e.g.*, New England, the Gulf States, the Lake region). Pay special attention to making forecasts of cold

WEATHER FORECASTING. 121

waves, of heavy rain or snowstorms, of high winds over the lakes or along the Atlantic coast, etc. When possible, obtain from the daily newspapers any particulars as to damage done by frost or gales, or concerning snow blockades, floods due to heavy rains, etc.

Fig. 53 summarizes what has thus far been learned as to the distribution of the various weather elements around a well-developed center of low pressure. The curved broken lines represent the *isotherms* (Chapter XIV). The solid concentric oval lines are the *isobars* (Chapter XI). The arrows represent the *winds*, the lengths of the arrows being roughly proportionate to the wind velocities (Chapter XII). The whole shaded area represents the region over which the sky is covered by heavy lower clouds. The smaller shaded area, within the larger, encloses the district over which rain or snow is falling (Chapter XVI). The lines running out in front of the cloudy area represent the light upper clouds (*cirrus* and *cirro-stratus*) which usually precede an area of low pressure.

Imagine this whole disturbance moving across the United States in a northeasterly direction, and imagine yourself at a station (1) directly in the path of the cyclone; (2) south of the track; and (3) north of the track. In the first case, as the disturbance moved on in its path, you would successively occupy the positions marked A, B, and C on the line AC, passing through the center of the cyclone. In the second case you would be first at D, then at E, and then at F. In the third case you would be at G, H, and J in succession. What changes of weather would you experience in each of these positions as the cyclone passed by you? Imagine yourself at some station halfway between the lines AC and DF. What weather changes would you have in that position with reference to the storm track? In what respects would these weather changes differ from those experienced along the line DF? Imagine your station halfway between the lines AC and GJ. What weather changes would you have there? How would these changes differ from those experienced along the line *GJ*?

It must be remembered that Fig. 53 is an ideal diagram. It represents conditions which are not to be expected in every cyclone which appears on our weather maps. If all cyclones were exactly alike in the weather conditions around them, weather forecasting would be a very easy task. But cyclones are not all alike — far from it. Some are well developed, with strong gradients, high winds, extended cloud areas, heavy precipitation, and decided temperature contrasts. Others are but poorly developed, with weak gradients, light winds, small temperature differences, and it may be without any precipitation whatever. Some cover immense districts of country; others are small and affect only a limited area. It therefore becomes necessary to examine the characteristics of each approaching cyclone, as shown on the daily weather map, very carefully. Notice whether it is accompanied by heavy rain or snow; whether its winds are violent; how far ahead of the center the cloudy area extends; how far behind the outer cloud limit the rain area begins; what is the position of the cloud and rain area with reference to the center, and other points of equal importance, and govern yourself, in making your forecast, according to the special features of each individual cyclone. Well-developed cyclones will be accompanied by marked weather changes. Weak cyclones will have their weather changes but faintly marked.

The distance of your station from the center of the cyclone is of great importance in determining what the weather conditions and changes shall be, as may easily be seen by examining Fig. 53. If the storm passes far to the north or far to the south of your station, you may notice none of its accompanying weather conditions, except, perhaps, a bank of clouds on your horizon. You may for a few hours be under the cloudy sky of some passing storm, and yet not be reached by its rainy area. The shifts in the wind may be marked and the wind velocities high, or the expected veering or backing may hardly be noticeable, owing to the weakness or the distance of the controlling cyclone.

Again, the rapidity with which weather conditions will change depends upon the rate of movement of the cyclone itself. The better developed the cyclone, the higher its velocity of progression, and the nearer its track lies to the station, the more emphatic and the more rapid are the weather changes it causes. On the other hand, the weaker the cyclone, the slower its rate of progression, and the further away its track, the less marked and the slower the weather changes. The probable track of a coming storm, and its probable rate of movement, therefore, need careful study if our forecasts are to be reliable.

There are many other obstacles in the way which combine to render weather forecasting extremely difficult. Some of these difficulties you will learn to overcome more or less successfully by the experience you will gain from a careful and persevering study of the daily weather maps; others, the best forecast officials of our Weather Bureau have not yet entirely overcome. The tracks followed by our cyclones vary more or less from month to month, and even if the average tracks for each month are known, individual cyclones may occur which absolutely disregard these tracks. While the average hourly velocity of cyclones is accurately known for the year and for each month, the movements of individual storms are often very capricious. They may move with a fairly uniform velocity throughout the time of their duration; they may suddenly and unexpectedly increase their rate of movement, or they may as suddenly come nearly to a standstill. The characteristics of cyclones vary in different portions of the country and at different times. Cyclones which have been accompanied by little precipitation on most of their journey are apt to give increased rain or snowfall as they near the Atlantic Ocean and Gulf of St. Lawrence. Cyclones which over one portion of the country were rainy, may give little or no precipitation in another portion. Cyclones and anticyclones are found to have considerable influence on one another, retarding or accelerating one another's advance, or changing one another's normal path of progression. While this mutual interaction is clearly seen, and may be successfully predicted in many cases, many other cases arise in which, under apparently similar conditions, the result is very different from the anticipation. Such are some of the difficulties with which weather forecasters have to contend, and which prevent the attainment of greater accuracy in weather prediction.

Part V. — Problems in Observational Meteorology.

CHAPTER XX.

TEMPERATURE.

The chief interest and value of the instrumental work in meteorology are to be found not only in the taking of the daily observations at stated hours, but in the working out of numerous simple problems, such as may readily be undertaken with the help of the instruments already described. Thus, the temperature of the air (obtained by the sling thermometer, supplemented by maximum and minimum thermometers, and by the thermograph if available) can be determined under a variety of conditions, *e.g.*, close to the ground, and at different heights above the ground; at different hours, by day and night; in different seasons; in sunshine and in shade; during wind and calms; in clear and cloudy weather; in woods and in the open; over bare ground, grass, snow, or ice; on hills and in valleys. Observations may also be made of the temperature of the ground and of a snow cover, at the surface and at slight depths beneath the surface, in different seasons and under different weather conditions. Among the problems which may be worked out by means of such observations as these are the following: —

A. **The Diurnal Range of Temperature under Different Conditions and at Different Heights above the Ground.** — Under the influence of the sun the regular normal variation of temperature during

24 hours is as follows: A gradual increase, with the increasing altitude of the sun, from sunrise until shortly after noon, and a gradual decrease, with decreasing altitude of the sun, from the maximum, shortly after noon, until the minimum, about sunrise. This variation is known as the *diurnal variation of temperature*. Curve *a* in Fig. 12 illustrates well the normal diurnal variation of temperature, as recorded by the thermograph during a period of clear, warm spring weather (April 27–30, 1889, Nashua, N. H.). The *diurnal range of temperature* is the difference between the maximum and minimum of the diurnal oscillation. The regular normal diurnal variation in temperature is often much interfered with by other controlling causes than the sun, *e.g.*, cyclonic winds, clouds, etc.

I. Study and compare the diurnal ranges of temperature as indicated by the maximum and minimum thermometers, or the thermograph, in the instrument shelter, in clear, fair, cloudy, and stormy weather, during winds and in calms, in different months. Summarize your results by grouping them according to the general weather conditions, and according to the months or seasons in which the observations were made. For example, group together and average the ranges observed on clear, calm days in winter; on similar days in early summer or autumn; on clear days with brisk northwest winds in winter; on similar days in early summer or autumn; on calm days with overcast sky in the different seasons; on stormy days with strong winds, etc. Study carefully the weather maps for the days on which your observations are made. Pay special attention to the relation between the diurnal ranges and the control exercised over these ranges by cyclones and anticyclones through their winds and general weather conditions.

II. Observations of diurnal ranges of temperature at different heights above the ground may be made by means of maximum and minimum thermometers fastened (temporarily) outside of the windows of different stories of the school or of some other building. These observations should be made out of windows facing north, and care should be taken to check, so far as possible, any draft from within the building out through the window during the taking of the observation. If a fire escape is provided on the building, the instruments may often be conveniently fastened to that.

Study the ranges under different conditions of wind and weather at various heights above the ground, and compare these results with those obtained under I. Notice the relations of all your results to the cyclonic and anticyclonic areas of the weather maps.

The diurnal range of temperature in the air over the open ocean from the equator to latitude 40° has been found to average only 2° to 3°. In southeastern California and the adjacent portion of Arizona the average diurnal temperature range in summer is 40° or 45°. Over other arid regions, such as the Sahara, Arabia, and the interior of Australia, the range also often amounts to 40°. Observations of temperature above the earth's surface, in the free air, made on mountains, in balloons, and by means of instruments elevated by kites, indicate very clearly that the diurnal range of temperature decreases with increasing elevation above sea level. The results obtained at Blue Hill Observatory, Massachusetts, by means of kites, show that the diurnal range of temperature almost disappears, on the average, at 3300 feet (1000 meters).

B. **Changes of Temperature in the Lower Air, and their Control by the Condition of the Ground, the Movement of the Air, and Other Factors.**— Determine the changes of the temperature in the lower air by making frequent readings of the ordinary thermometer in the instrument shelter, of the sling thermometer, or by an examination of the thermograph record. Group these changes, as in Problem *A*, so far as possible according to the weather conditions under which they occurred, and try to classify the kinds of change roughly into types. Study the control of these various types by the wind and other weather conditions accompanying them, as illustrated on the daily weather maps. The control exercised by different conditions of the earth's surface may be studied by means of observations made with the sling thermometer over different surfaces, such as grass, bare ground, snow, etc.

Examples of temperature changes in the lower air, under different conditions of weather, recorded on the thermograph, are given in Fig. 12, and are briefly referred to their causes in the text accompanying that figure.

C. **Vertical Distribution of Temperature in the Atmosphere.**— The vertical distribution of temperature in the lower air may be studied by having ordinary thermometers or thermographs exposed at different heights above the ground, *e.g.*, close to the surface ; in an instrument shelter ; out of windows on successive stories of some high building ; and on the roof of the building. They may also, in cases where there is a hill in the neighborhood, be exposed in a valley at the bottom of the hill and at successive elevations up the side of the hill. It is, however, usually much simpler, as well as more practicable, to take these temperature readings by means of the sling thermometer. In the case of observations made out of the windows of a building, one observer can take the readings at different elevations in succession. When the observations are made at different altitudes on the side of a hill, it is best to have the coöperation of several observers, who shall all read their thermometers at the same moment of time. The results obtained in the previous problems (*A* and *B*) may, of course, also be utilized in studying the vertical distribution of temperature in the atmosphere.

Study the vertical distribution of temperature in the lower air under various conditions of weather and season ; at various hours of the day, and with varying conditions of surface cover. Make your observations systematically, at regular hours, so that the results may be comparable. Group together observations made under similar conditions of weather, season, time, and surface cover. Determine the average vertical distribution of temperature in the different cases. Note especially any seeming peculiarities or irregularities in this distribution at certain times. Study carefully, as in the previous problems, the relation of the different types of temperature distribution in the atmosphere to the weather conditions as shown on the daily weather maps.

Observations made in different parts of the world, on mountains and in balloons, have shown that on the average the temperature decreases from the earth's surface upwards at the rate of about 1° in 300 feet of ascent. The rate of vertical decrease of temperature is known in meteorology as the *vertical temperature gradient*. When it happens that there is for a time an *increase in temperature upwards* from the earth's surface, the condition is known as an *inversion of temperature*.

As a result of the decrease of temperature with increasing altitude above sea level, the tops of many high mountains even in the Torrid Zone are always covered with snow, while no snow can ever fall at their bases, owing to the high temperatures which prevail there. Balloons sent up without aëronauts, but with self-recording instruments, have given us temperatures of $-90°$ at a height of 10 miles above the earth's surface. On Dec. 4, 1896, Berson reached a height of 30,000 feet and noted a temperature of $-52°$. Inversions of temperature are quite common, especially during the clear cold spells of winter. Under such conditions the tops and sides of hills and mountains are often much warmer than the valley bottoms at their bases. A good example of an inversion of temperature occurred in New Hampshire on Dec. 27, 1884. The pressure was above the normal, the sky clear and the wind light. The observer on the summit of Mt. Washington reported a temperature of $+16°$ on the morning of that day, while the thermometers on the neighboring lowlands gave readings of from $-10°$ to $-24°$. In Switzerland, the villages and cottages are generally built on the mountain sides and not down in the valley bottoms, experience having taught the natives that the greatest cold is found at the lower levels.

CHAPTER XXI.

WINDS.

THE determination of the direction of the wind (by means of the wind vane) and of its velocity (by means of the anemometer, or by estimating its strength) at different hours, under different conditions of weather and in different seasons, leads to a number of problems. The following simple investigations may readily be undertaken in schools : —

A. **The Diurnal Variation in Wind Velocity in Fair Weather.** — Observe and record the velocity of the wind (either estimated or registered by the anemometer) every hour, or as often as possible, on clear or fair days in different months. Can you discover any regular change in the velocities during the day? If so, what is the change? Does the season seem to have any control over the results obtained? Examine the daily weather maps in connection with your observations and determine the effect that different weather conditions have upon the diurnal variation in wind velocity.

The diurnal variation in wind velocity over the open ocean is so slight as hardly to be noticeable. Over the land, the daytime winds are commonly strongest in arid regions. Traveling across the desert often becomes extremely disagreeable, owing to the clouds of dust which these winds sweep up from the surface.

B. **The Variations in Direction and Velocity due to Cyclones and Anticyclones.** — Record the direction and velocity of the wind at your station at frequent intervals during the passage of a considerable number of cyclones and anticyclones. Enter your observations in some form of table so that they may be readily examined. (See p. 113.) Note the character of the changes that occur, classifying them into types, so far as possible. Study the control of wind directions and velocities by the special features of the individual cyclones and anticyclones as shown on the daily weather maps. How are the different types of change in direction and velocity affected by the tracks of cyclones and anticyclones? By their velocity of progression? By the arrangement of isobars around them? By the height of the barometer at the center? By the season in which the cyclones and anticyclones occur?

Frequent changes in the direction and velocity of our winds are one great characteristic of the Temperate Zones, especially in winter. The continuous procession of cyclones and anticyclones across the United States involves continuous shifts of wind. Over much of the earth's surface, however, the regularity and constancy of the winds are the distinguishing feature of the climate. Over a considerable part of the belts blown over by the northeast and southeast trades, roughly between latitude 30° N. and S. and the equator, the winds keep very nearly the same direction and the same velocity day after day and month after month. Thus the trades are of great benefit to commerce. Sailing ships may travel for days in the trade wind belts without having their sails shifted at all, with a fair wind all the time carrying them rapidly on to their destination.

C. **The Occurrence and Characteristics of Local Winds, such as Mountain and Valley and Land and Sea Breezes.** — If the observer happens to be living in or near the mouth of a valley or on a mountain side, opportunity may be given for the observation of the local winds down the mountain sides and down the valley at night, and up the valley and the mountain sides by day, known as mountain and valley breezes. Keep a record of wind direction and velocity during the day, and especially during the morning and evening hours. Notice any marked changes in direction, and the relation of these changes to the time of day. Does the velocity of the daytime up-cast breeze show any systematic variation during the day? Study the relation of mountain and valley breezes to the general weather conditions shown on the weather maps. How are these breezes affected by season? By the presence of a cyclone over the region? Of an anticyclone? By the state of the sky?

If near the seacoast (*i.e.*, within 10 or 15 miles), an interesting study may be made of local land and sea breezes. The sea breeze is a wind from the ocean onshore, while the land breeze blows offshore. These breezes occur only in the warmer months. Take frequent observations during the day, as in the case of mountain and valley winds, noting especially any changes in direction and velocity, and the relation of these changes to the time of day. Study also the control exercised by the prevailing weather conditions over the occurrence and the strength of development of the land and sea breezes.

This problem may be considerably extended by adding temperature observations to the simpler record of wind direction and velocity.

In some of the Swiss valleys the mountain and valley breezes are such regular daily weather phenomena that it has become a weather proverb that a failure of the daily change in wind direction indicates a change of weather. Special names are often given to these breezes where they are well marked. In a part of the Tyrol sailing boats go up the lakes by day with the valley breeze, and sail back at night with the mountain breeze. It is therefore unnecessary for the boats to be rowed either way. Land and sea breezes, although an unimportant climatic feature in these northern latitudes, are often of the highest importance in the Torrid Zone. The fresh pure sea breeze from over the ocean makes it possible for Europeans to live in many tropical climates where otherwise they would not keep their health. The land breeze, on the other hand, is apt to be an unhealthy wind in the tropics, especially when it blows off of swampy land.

CHAPTER XXII.

HUMIDITY, DEW, AND FROST.

THE humidity of the air, as determined by the wet and dry bulb thermometers or the sling psychrometer, and the occurrence or absence of dew or frost, should be studied together.

Observations should be made at different hours, in different kinds of weather, and in different seasons. From such observations the following problems may be solved : — .

A. **Diurnal Variation of Relative Humidity under Different Conditions.** — Readings of the wet and dry bulb thermometers in the instrument shelter, or of the sling psychrometer, several times during the day, will furnish data for determining the diurnal variation of relative humidity. Classify your observations according to the weather conditions under which they were made, and by months or seasons. Summarize the results of your investigation, paying special attention to the relation between the diurnal variation of relative humidity and the temperature.

The variations of relative humidity are generally the reverse of those of absolute humidity. In the case of the latter the average diurnal variations are small. The fluctuations in the relative humidity during the day on the northwestern coast of Europe amount to about 7% in December and 17% in August, while in central Asia they average about 25% in winter and 50% in summer.

B. **Relation of Relative Humidity to the Direction of the Wind.** — Observations by means of the wet and dry bulb thermometers in the shelter, or by means of the sling psychrometer, supplemented by records of wind direction, will furnish data for the solution of this problem. Tabulate your observations according to wind directions and seasons. Determine the characteristics of the different winds as to their relative humidities. Consider the control of these winds and humidity conditions by cyclones and anticyclones.

The warm wave, or sirocco, in front of our winter cyclones in the eastern United States is a damp, disagreeable, irritating wind. In summer, the sirocco is usually dry, and during the prevalence of such winds we have our hottest spells, when sunstrokes are not uncommon. In southern Italy the sirocco has the same position with reference to the controlling cyclone. There the wind is often so dry as seriously to injure vegetation. The cold wave, on the rear of our winter cyclones, with its low temperature and dry air, often comes as a refreshing change after the enervating warmth of the preceding sirocco. Our feelings of bodily comfort or discomfort are thus in a large measure dependent upon the humidity and the movement of the air.

C. **The Formation of Dew.** — The formation of dew is to be studied from the following points of view, viz., as dependent upon : *a*, the temperature and the humidity of the air ; *b*, the exposure and condition of the ground ; *c*, the state of the sky ; and *d*, the movement of the air. The occurrence of dew on any night, as well as the amount, whether large or small, can readily be ascertained by inspection. Observe the conditions of temperature, humidity, cloudiness, and wind direction and velocity, as in previous exercises. Pay special attention to the state of the sky, the wind movement, and the vertical distribution of temperature near the ground. Under heading *b* (exposure and condition of the ground) make observations of the amounts of dew formed on hilltops, hillsides, and in valleys ; on different kinds of surface covering, as grass, leaves, pavements, etc., and over different kinds of soil. Classify the results in accordance with the conditions under which the observations were made. Compare the results and draw your conclusions from this study. Practise making predictions of the formation of dew in different places and under different weather conditions.

Over the greater portion of the earth's surface the amount of dew which is deposited is very small. It has been estimated that in Great Britain the total annual amount would measure only an inch and a half in depth ; and in central Europe the depth is given as hardly one inch. In some parts of the Torrid Zone, on the other hand, dew is deposited in much larger quantities. According to Humboldt, the traveler through some of the South American forests often finds what seems to be a heavy shower falling under the trees, while the sky is perfectly clear overhead. In this case dew is formed on the tops of the tree in sufficiently large quantities to give a shower underneath. It is reported that on the Guinea coast of Africa the dew sometimes runs off the roofs of the huts like rain. In many dry regions the dew is an important agency in keeping the plants alive.

D. **The Formation of Frost.** — The formation of frost is to be studied in the same way as that suggested in the case of dew, *i.e.*, as dependent upon: *a*, the temperature and the humidity of the air; *b*, the exposure and condition of the ground; *c*, the state of the sky; and *d*, the movement of the air. Frosts are usually classified as *light* or *heavy*. The words *killing frost* are also used. Study the weather and surface conditions which are most favorable to the formation of frost. Pay special attention to the relation of frost and inversions of temperature; to the frequency of frost on open or sheltered surfaces; on hills or in valleys, and on the lower and upper branches of trees and shrubs. Determine, as well as you can, the weather conditions which precede light or heavy frosts, and make predictions of coming frosts, when the conditions warrant them.

Our Weather Bureau gives much attention to the prediction of frosts and to the prompt and widespread distribution of frost warnings. Growing crops and fruits are often seriously injured by frosts, and farmers are naturally anxious to have as early warning as possible of their occurrence. Various methods of protecting crops and trees against frost are used. The method most commonly employed consists in the building of fires of brush or other inflammable material on the windward side of the field or the orchard when a frost is expected. The smoke from the fire is blown to leeward across the field, and acts as an artificial cloud, affording protection to the vegetation underneath. Such fires are known as *smudges*.

CHAPTER XXIII.

CLOUDS AND UPPER AIR CURRENTS.

ATTENTIVE observation of clouds will soon lead to a familiarity with their common type forms. A series of cloud views,[1] with accompanying descriptive accounts, will teach the names of the clouds and give definiteness to the record. The directions of movement of clouds are determined by means of the nephoscope. Cloud observations should be made at different hours, in different weather conditions, and in different seasons. The following problems are concerned with clouds and upper air currents : —

A. **The Typical Cloud Forms and their Changes.** — Note carefully the characteristic forms assumed by clouds ; their mode of occurrence, whether in single clots, or in groups, in lines, or all over the sky ; their changes in form and in mode of occurrence. Classify and summarize your results. Compare the clouds of the warm months with those of the cold months.

Observations have shown that clouds have certain definite characteristic forms which are substantially the same in all parts of the world. This fact makes it possible to give names to the different typical forms, and these names are used by observers the world over. Hence cloud observations, wherever made, are comparable. The first classification of clouds was proposed by Luke Howard, in 1803. The classification at present in use is known as the *International Classification*, and was adopted by the International Meteorological Congress in 1896.

B. **The Prevailing Direction of Cloud Movements.** — The use of the nephoscope is necessary in the accurate determination of cloud movements. Study the prevailing directions of movement of the clouds, by means of frequent observations with the nephoscope, in different weather conditions. Separate the

[1] See *Hydrographic Office Cloud Types*, Appendix B.

upper and lower clouds in this study. Summarize your results according to the weather conditions and the kinds of clouds.

C. **Correlation of Cloud Form and Movement with Surface Winds, with Cyclones and Anticyclones, and with Weather Changes.** — The results obtained in the working out of the two preceding problems may be used in the present problem. Tabulate your observations of cloud forms with reference to the wind directions which prevailed at the time of making the observations. Do the same with the directions of cloud movement. Determine the relation between surface winds and cloud types, and between surface winds and the direction of the upper air currents, as shown by the movements of the upper clouds. Study the control exercised by cyclones and anticyclones over cloud forms and over the direction of the upper air currents.

D. **The Use of Clouds as Weather Prognostics.** — Attentive observation of the forms and changes of clouds, and of the accompanying and following weather changes, will lead to the association of certain clouds with certain coming weather conditions. Make your cloud observations carefully, taking full notes at the time of observation. Give special attention to the weather conditions that follow. Continue this investigation through as long a period as possible, until you have gathered a considerable body of fact to serve as a basis, and then frame a set of simple rules for forecasting fair or stormy weather on the basis of the forms and changes of the clouds. Such local observations as these may be employed as a help in making forecasts from the daily weather maps.

Clouds were used as weather prognostics long before meteorological observations and weather maps were thought of. To-day sailors and farmers still look to the clouds to give them warning of approaching storms. Many of our common weather proverbs are based on the use of clouds as weather prognostics.

CHAPTER XXIV.

PRECIPITATION.

THE special study of various problems connected with precipitation involves detailed observations of the amount and rate of precipitation of various kinds, measured by the rain gauge during storms in different seasons. These observations of precipitation should, of course, be supplemented by the usual record of the other weather elements. The following problems are suggested : —

A. *The relation of precipitation in general to the other weather elements, and to cyclones and anticyclones.*

B. *The conditions under which special forms of precipitation (rain, snow, sleet, hail, frozen rain) occur.*

C. *The conditions associated with light and heavy, brief and prolonged, local and general rainfall.*

These problems are studied by means of a careful comparison of full weather records with the daily weather maps during a considerable period of time.

Rain is the most common form of precipitation the world over, although snow falls over large portions of both hemispheres. In the Arctic and Antarctic zones almost all the precipitation, which is small in amount, comes in the form of snow. In southern Europe snow falls at sea level during the winter as far south as 36° north latitude on the average. In eastern Asia snow occasionally falls as far south as 23° north latitude. The mean annual rainfall varies greatly in different parts of the world. In desert regions it is practically nothing. At Cherrapunjee, in India, it reaches 493 inches, or over 40 feet. A fall of 40.8 inches in a single day occurred at this station on June 14, 1876. In the United States, Upper Mattole, Cal., had an extraordinary monthly rainfall of 41.63 inches in January, 1888. An excessive daily rainfall of 8 inches occurred at Syracuse, N. Y., on June 8, 1876. At Washington, D. C., 2.34

inches fell in 37 minutes on June 27, 1881. A sudden and very heavy fall of rain occurred at Palmetto, Nevada, in August, 1890. A rain gauge which was not exposed to the full intensity of the storm caught 8.80 inches of water in one hour. In August, 1891, an observer at Campo, Cal., measured 11.5 inches as the rainfall in one hour from one very heavy downpour, and from a portion of a second storm.

CHAPTER XXV.

PRESSURE.

THE variations of atmospheric pressure, although insensible to non-instrumental observation, are so intimately connected with atmospheric processes that they deserve careful attention. Their observation leads to several problems.

A. **The Decrease of Pressure with Height, as between Valley and Hill, or between the Base and Top of a Building.** — Make these observations with the mercurial barometer, if possible. Note the air temperatures at the two levels at which the barometer readings are made. Determine the heights of hill or building by means of the following rule: Multiply by 9 the difference in barometrical readings at the two stations, given in hundredths of an inch, and the result will be approximately the difference in height between the stations in feet. A more accurate result may be reached by means of the following rule: The difference of level in feet is equal to the difference of the pressures in inches divided by their sum, and multiplied by the number 55,761, when the mean of the air temperatures of the two places is 60°. If the mean temperature is above 60°, the multiplier must be increased by 117 for every degree by which the mean exceeds 60°; if less than 60°, the multiplier must be decreased in the same way. For example, if the lower station has a pressure of 30.00 inches and a temperature of 62°, and the upper station has 29.00 inches and 58° respectively, the difference of level between the two will be

$$\frac{30.00 - 29.00}{30.00 + 29.00} \times 55{,}761 = 945 \text{ feet.}$$

If the lower values are 30.15 inches and 65°, while the upper values are 28.67 inches and 59°, then the formula becomes

$$\frac{30.15 - 28.67}{30.15 + 28.67} \times [55{,}761 + (2 \times 117)] = 1409 \text{ feet.}$$

The determination of heights by means of the barometer depends upon the fact that the rate of decrease of pressure upwards is known. As the weight of a column of air of a given height varies with the temperature of the air, it is necessary, in accurate work of this sort, to know the air temperatures at both the lower and upper stations at the time of observation. From these temperatures the mean temperature of the air column between the two stations may be determined. Tables have been published which facilitate the reductions in this work. The heights of mountains are usually determined, in the first instance, by means of barometric observations, carried out by scientific expeditions or by travelers that have been able to reach their summits. More accurate measurements are later made, when possible, by means of trigonometrical methods.

B. **The Diurnal and Cyclonic Variation of Pressure in Different Seasons.** — This problem is satisfactorily solved only by a study of the curves traced by the barograph, or by plotting, as a curve, hourly or half-hourly readings of the mercurial barometer. The *diurnal variation of the barometer* is the name given to a slight double oscillation of pressure, with two maxima and two minima occurring during the 24 hours. This oscillation is in some way, not yet understood, connected with the diurnal variation in temperature. It is most marked in the tropics and diminishes towards the poles. Fig. 15 illustrates, in the May curve, the diurnal variation of the barometer at

Cambridge, Mass., during a spell of fair spring weather, May 18-22, 1887. The maxima are marked by + and the minima by 0. The *cyclonic variation of pressure* is the name given to those irregular changes in pressure which are caused by the passage of cyclones and anticyclones. The second curve in Fig. 15 shows the cyclonic variations in pressure recorded by the barograph at Cambridge, Mass., during a spell of stormy weather, Feb. 23-28, 1887. These curves serve as good illustrations of these two kinds of pressure variations.

Study your barograph tracings, or your barometer readings, as illustrating diurnal or cyclonic variations of pressure. Note the character and the amount of the diurnal and cyclonic variations, and their dependence on seasons.

Over the greater part of the Torrid Zone the diurnal variation of the barometer is remarkably distinct and regular. Humboldt first called attention to the fact that in those latitudes the time of day may be told within about 15 minutes if the height of the barometer is known.

C. **The Relation of Local Pressure Changes to Cyclones and Anticyclones, and thus to Weather Changes.** — Make a detailed study of the relation of the local pressure changes at your station, as shown by the barograph curves, or by frequent readings of the mercurial or aneroid barometers, to the passage of cyclones and anticyclones, and to their accompanying weather changes. Classify the simple types of pressure change, so far as possible, together with the general weather conditions that usually accompany these types. Apply the knowledge of local weather changes thus gained when you make your forecast on the basis of the daily weather maps.

CHAPTER XXVI.

METEOROLOGICAL TABLES.

THE tables which follow are those which are now in use by the United States Weather Bureau. They were first published in the *Instructions for Voluntary Observers* issued in 1892, and were reprinted in 1897. The following instructions will be found of service in the use of the tables: —

TABLE I. — DEW-POINT.

The figures in heavy type, arranged in vertical columns at each side of the page, are the *air temperatures* in degrees Fahrenheit, as recorded by the dry-bulb thermometer. The figures in heavy type, running across the page, denote the *differences*, in degrees and tenths of degrees, *between the dry-bulb and wet-bulb readings*, or, technically, the *depression of the wet-bulb thermometer*. The figures in the vertical columns denote the *dew-points*. Make your observation of the wet and dry bulb thermometers and note the difference between the two readings. Find, in the vertical columns of heavy type, the temperature corresponding to your dry-bulb reading, or the nearest temperature to that. Then look along the horizontal lines of figures in heavy type for the figure which corresponds exactly, or most nearly, with the difference between your wet and dry bulb readings. Look down the vertical column under this latter figure until you reach the horizontal line corresponding to your dry-bulb reading. At this point the figures in the vertical column give the dew-point of the air at the time of your observation.

Example: Air Temperature (dry bulb), 47°; Wet Bulb, 44°; Difference, 3°. On page 148 will be found the table containing both 47° (dry bulb) and 3° (depression of the dew-point). In the twenty-eighth line of this table and in the seventh column will be found the dew-point, viz., 41°.

Example: Air Temperature, 61.5°; Wet Bulb, 55.5°; Difference, 6°.

In this case 61.5° is not found in the vertical columns of dry-bulb readings, but 61° and 62° are found. The dew-point, with a difference between wet and dry bulb readings of 6°, for an air temperature of 61°, is 50°; for an air temperature of 62°, it is 52°. Evidently, then, for an air temperature of 61.5° the dew-point will be 51°, *i.e.*, halfway between 50° and 52°. This method of determining dew-points at air temperatures or with depressions of the wet-bulb thermometer which are not given exactly in the tables, is known as *interpolation*.

Example: Air Temperature, 93°; Wet Bulb, 90.5°; Difference, 2.5°. Our table gives no dew-points for wet-bulb depressions of 2.5°, with air temperature 93°, but we find (on page 152) that for air temperature 93° and depression of wet bulb of 2°, the dew-point is 91°, while for a wet-bulb depression of 3°, the dew-point is 89°. By the method of interpolation we can readily determine the dew-point in the special case under consideration as 90°, *i.e.*, halfway between 89° and 91°.

TABLE II.— RELATIVE HUMIDITY.

The general plan of this table is the same as that of Table I. The figures in the vertical columns are the relative humidities (in percentages) corresponding to the different readings of the wet and dry bulb thermometers.

TABLE III.— REDUCTION OF BAROMETER TO 32°.

The figures in heavy type, arranged in vertical columns at the left of the page, refer to the temperature in degrees Fahrenheit, as indicated by the attached thermometer. The figures in heavy type, running across the top of the page, are the barometer readings in inches and tenths. Make a reading of the attached thermometer and of the barometer. Find in the vertical column the temperature corresponding to the reading of the attached thermometer, and in the horizontal line of heavy figures the reading corresponding to the height of the barometer. The decimal in the vertical column, under the appropriate barometer reading, and in the same horizontal line with the appropriate thermometer reading, is to be subtracted from the height of the barometer as observed, thus correcting the reading to freezing. When the attached thermometer reads below 28°, the correction is additive.

Example: Attached Thermometer, 69°; Barometer, 30.00 inches; Correction, − .110; Corrected reading, 29.890 inches.

Example: Attached Thermometer, 73°; Barometer, 29.75 inches; Correction = ?

We do not find any column corresponding to a barometer reading of 29.75 inches. We do find, however, that with a barometer reading of 29.50, and an attached thermometer reading of 73, the correction is − .118 inch, and with a barometer reading of 30.00, the correction is −.120. By interpolating, as in the case of the humidity table above, we find the correction for a barometer reading of 29.75 inches, and an attached thermometer reading of 73°. The correction is − .119, and the corrected reading is 29.75 − .119 = 29.63 inches.

TABLE IV.— REDUCTION OF BAROMETER TO SEA LEVEL.

The figures in heavy type, in the left-hand vertical columns, are the heights, in feet, of the barometer above sea level. The figures in heavy type at the top of the columns, running across the page, are the readings of the ordinary thermometer. The numbers of inches and hundredths of inches to be subtracted from the barometer reading (corrected for temperature by Table

III), for the different heights above sea level, are given in the vertical columns.

The altitude above sea level of the city or town at which the observation is made should be ascertained as accurately as possible from some recognized authority, as, *e.g.*, from a railroad survey; from Government measurements, or from some engineer's office. The correction to be made is determined by a simple inspection of the table or by the method of interpolation.

Example: Altitude of Barometer above sea level, 840 feet; Temperature of the air, 40°; Correction, +.931 inch.

Example: Altitude of Barometer above sea level, 205 feet; Temperature of the air, 45°; Correction = ?

Here 205 feet and 45° are neither of them found in the table. Hence a double interpolation is necessary. For 200 feet and 40° the correction is + .224 inch. For 200 feet and 50° the correction is + .220 inch. Hence for 200 feet and 45° the correction is + .222 inch. For 210 feet and 40° the correction is + .235 inch. For 210 feet and 50° the correction is + .231 inch. Hence for 210 feet and 45° the correction is + .233 inch. Now for 205 feet we should have a correction midway between + .235 inch and + .233 inch or + .234 inch.

Table I. — Temperature of the Dew-Point, in Degrees Fahrenheit.

t (Dry ther.)	Difference between the dry and wet thermometers (t—t').												t (Dry ther.)
	0°.2	0°.4	0°.6	0°.8	1°.0	1°.2	1°.4	1°.6	1°.8	2°.0	2°.2	2°.4	2°.6	
−40	−52													−40
−39	−50													−39
−38	−49													−38
−37	−48													−37
−36	−46													−36
−35	−44													−35
−34	−43	−58												−34
−33	−42	−55												−33
−32	−40	−52												−32
−31	−38	−49												−31
−30	−36	−47												−30
−29	−35	−44												−29
−28	−33	−42	−56											−28
−27	−32	−40	−52											−27
−26	−30	−37	−48											−26
−25	−29	−35	−45											−25
−24	−28	−34	−43	−58										−24
−23	−27	−32	−40	−53										−23
−22	−26	−30	−37	−49										−22
−21	−25	−29	−35	−45										−21
−20	−23	−28	−33	−41	−55									−20
−19	−22	−26	−31	−38	−50									−19
−18	−21	−25	−29	−35	−45									−18
−17	−20	−23	−27	−32	−41	−55								−17
−16	−19	−22	−26	−30	−37	−49								−16
−15	−17	−20	−24	−28	−34	−44								−15
−14	−16	−19	−22	−26	−31	−39	−52							−14
−13	−15	−18	−21	−25	−29	−35	−46							−13
−12	−14	−17	−20	−23	−27	−32	−41	−55						−12
−11	−13	−16	−18	−21	−25	−30	−36	−48						−11
−10	−12	−14	−17	−20	−23	−27	−33	−42	−58					−10
−9	−11	−13	−15	−18	−21	−25	−30	−37	−48					−9
−8	−10	−12	−14	−17	−20	−23	−27	−33	−42	−58				−8
−7	−9	−11	−13	−15	−18	−21	−25	−30	−36	−48				−7
−6	−8	−10	−12	−14	−16	−19	−23	−27	−32	−41	−56			−6
−5	−7	−8	−10	−12	−15	−17	−21	−24	−29	−35	−47			−5
−4	−6	−7	−9	−11	−13	−16	−19	−22	−26	−31	−39	−54		−4
−3	−4	−6	−8	−10	−12	−14	−17	−20	−23	−28	−33	−44		−3
−2	−3	−5	−6	−8	−10	−12	−15	−18	−21	−24	−29	−36	−48	−2
−1	−2	−4	−5	−7	−9	−11	−13	−16	−18	−22	−26	−31	−39	−1
0	−1	−3	−4	−6	−7	−9	−11	−14	−16	−19	−23	−27	−33	0
+1	−0	−2	−3	−4	−6	−8	−10	−12	−14	−17	−20	−24	−28	+1
2	+1	−1	−2	−3	−5	−6	−8	−10	−12	−15	−17	−21	−25	2
3	2	+1	−1	−2	−3	−5	−7	−8	−10	−13	−15	−18	−21	3
4	3	+2	0	−1	−2	−4	−5	−7	−9	−11	−13	−16	−19	4
5	4	3	+1	0	−1	−2	−4	−5	−7	−9	−11	−14	−16	5
6	5	4	3	+1	0	−1	−3	−4	−6	−7	−9	−12	−14	6
7	6	5	4	3	+1	0	−1	−3	−4	−6	−8	−10	−12	7
8	7	6	5	4	3	+1	0	−1	−3	−4	−6	−8	−10	8
9	8	7	6	5	4	3	+1	0	−1	−3	−4	−6	−8	9
10	9	8	7	6	5	4	3	+1	0	−1	−3	−4	−6	10
11	10	9	8	7	6	5	4	3	+2	0	−1	−3	−4	11
12	11	10	9	8	7	5	4	3	+2	+2	0	−1	−2	12
13	12	11	11	10	9	8	7	6	5	3	+2	+1	−1	13
14	13	12	12	11	10	9	8	7	6	5	4	2	+1	14
15	14	13	13	12	11	10	9	8	7	6	5	4	3	15
16	15	15	14	13	12	11	10	10	9	8	7	5	4	16
17	16	16	15	14	13	12	12	11	10	9	8	7	6	17
18	17	17	16	15	14	14	13	12	11	10	9	8	7	18
19	18	18	17	16	15	15	14	13	12	11	10	9	9	19
20	19	19	18	17	17	16	15	14	13	13	12	11	10	20
t.	0°.2	0°.4	0°.6	0°.8	1°.0	1°.2	1°.4	1°.6	1°.8	2°.0	2°.2	2°.4	2°.6	t.

METEOROLOGICAL TABLES. 147

TABLE I.—TEMPERATURE OF THE DEW-POINT, IN DEGREES FAHRENHEIT.

t (Dry ther.)	Difference between the dry and wet thermometers ($t-t'$).									t (Dry ther.)
	2°.6	2°.8	3°.0	3°.2	3°.4	3°.6	3°.8	4°.0	4°.2	4°.4	
− 2	−48										− 2
− 1	−39	−54									− 1
0	−33	−43									0
+ 1	−28	−35	−46								+ 1
2	−25	−30	−37	−50							2
3	−21	−26	−31	−39	−54						3
4	−19	−22	−27	−32	−42	−60					4
5	−16	−19	−23	−28	−34	−45					5
6	−14	−17	−20	−24	−29	−35	−47				6
7	−12	−14	−17	−20	−24	−29	−37	−50			7
8	−10	−12	−15	−17	−21	−25	−30	−38	−53		8
9	− 8	−10	−12	−15	−18	−21	−25	−31	−39	−55	9
10	− 6	− 8	−10	−12	−15	−18	−21	−26	−31	−40	10
11	− 4	− 6	− 8	−10	−12	−15	−18	−21	−26	−31	11
12	− 2	− 4	− 6	− 8	−10	−12	−15	−18	−21	−26	12
13	− 1	− 2	− 4	− 5	− 7	− 9	−12	−14	−17	−21	13
14	+ 1	0	− 2	− 3	− 5	− 7	− 9	−11	−14	−17	14
15	3	+ 1	0	− 2	− 3	− 5	− 7	− 9	−11	−14	15
16	4	3	+ 2	0	− 1	− 3	− 4	− 6	− 8	−10	16
17	6	5	3	+ 2	+ 1	− 1	− 2	− 4	− 6	− 8	17
18	7	6	5	4	2	+ 1	0	− 2	− 3	− 5	18
19	9	8	7	5	4	3	+ 1	0	− 1	− 3	19
20	10	9	8	7	6	5	3	+ 2	+ 1	− 1	20
t.	2°.6	2°.8	3°.0	3°.2	3°.4	3°.6	3°.8	4°.0	4°.2	4°.4	t.

t.	4°.6	4°.8	5°.0	5°.2	5°.4	5°.6	5°.8	6°.0	6°.2	6°.4	t.
8											8
9											9
10	−57										10
11	−41	−60									11
12	−31	−41	−59								12
13	−25	−31	−40	−58							13
14	−20	−25	−30	−39	−56						14
15	−16	−20	−24	−30	−38	−53					15
16	−13	−16	−19	−23	−28	−36	−50				16
17	−10	−12	−15	−18	−22	−27	−34	−47			17
18	− 7	− 9	−12	−14	−17	−21	−26	−32	−44		18
19	− 5	− 7	− 9	−11	−14	−17	−20	−25	−30	−40	19
20	− 2	− 4	− 6	− 8	−10	−13	−16	−19	−23	−29	20
t.	4°.6	4°.8	5°.0	5°.2	5°.4	5°.6	5°.8	6°.0	6°.2	6°.4	t.

TABLE I. — TEMPERATURE OF THE DEW-POINT, IN DEGREES FAHRENHEIT.

t (Dry ther.)	Difference between the dry and wet thermometers ($t - t'$).											t (Dry ther.)
	0°.5	1°.0	1°.5	2°.0	2°.5	3°.0	3°.5	4°.0	4°.5	5°.0	5°.5	6°.0	
20	18	17	15	13	10	8	5	2	−2	−6	−12	−19	20
21	19	18	16	14	12	9	7	4	0	−4	− 8	−15	21
22	20	19	17	15	13	11	8	6	+2	−1	− 6	−11	22
23	22	20	18	16	14	12	10	7	4	+1	− 3	− 8	23
24	23	21	19	18	16	14	11	9	6	3	− 1	− 5	24
25	24	22	21	19	17	15	13	11	8	5	+ 2	− 2	25
26	25	23	22	20	18	16	14	12	10	7	4	0	26
27	26	24	23	21	20	18	16	14	11	9	6	+ 3	27
28	27	25	24	22	21	19	17	15	13	11	8	5	28
29	28	26	25	24	22	20	19	17	14	12	10	7	29
30	29	27	26	25	23	22	20	18	16	14	11	9	30
31	30	29	27	26	24	23	21	19	18	15	13	11	31
32	31	30	28	27	26	24	22	21	19	17	15	13	32
33	31	31	29	28	26	25	23	22	19	18	16	14	33
34	32	32	30	29	27	26	24	24	21	20	18	16	34
35	33	32	31	30	29	28	26	25	23	22	20	18	35
36	35	34	32	31	30	29	27	26	24	23	21	19	36
37	36	35	33	32	31	30	28	27	26	24	22	21	37
38	37	36	34	33	32	31	30	28	27	26	24	22	38
39	38	37	35	34	33	32	30	29	28	27	25	24	39
40	39	38	36	35	34	33	31	30	29	28	26	25	40
41	40	39	37	36	35	34	32	32	30	29	28	26	41
42	41	40	39	38	36	35	34	33	31	30	29	27	42
43	42	41	40	39	37	36	35	34	32	31	30	29	43
44	43	42	41	40	38	37	36	35	33	32	31	30	44
45	44	43	42	41	40	39	37	36	34	33	32	31	45
46	45	44	43	42	41	40	38	37	36	35	33	32	46
47	46	45	44	43	42	41	40	39	37	36	34	33	47
48	47	46	45	44	43	42	41	40	38	37	36	35	48
49	48	47	46	45	44	43	42	41	39	38	37	36	49
50	49	48	47	46	45	44	43	42	41	40	38	37	50
51	50	49	48	47	46	45	44	43	42	41	39	38	51
52	51	50	49	48	47	46	45	44	43	42	41	40	52
53	52	51	50	49	48	47	46	45	44	43	42	41	53
54	53	52	51	50	50	49	47	46	45	44	43	42	54
55	54	53	53	52	51	50	49	48	47	46	44	43	55
56	55	54	54	53	52	51	50	49	48	47	45	44	56
57	56	55	55	54	53	52	51	50	49	48	47	46	57
58	57	56	56	55	54	53	52	51	50	49	48	47	58
59	58	57	57	56	55	54	53	52	51	50	49	48	59
60	59	58	58	57	56	55	54	53	52	51	50	49	60
61	60	59	59	58	57	56	55	54	53	52	51	50	61
62	61	60	60	59	58	57	56	55	54	53	52	52	62
63	62	61	61	60	59	58	57	56	55	55	54	53	63
64	63	62	62	61	60	59	58	57	56	56	55	54	64
65	64	63	63	62	61	60	59	59	58	57	56	55	65
66	65	64	64	63	62	61	60	60	59	58	57	56	66
67	67	66	65	64	63	62	61	61	60	59	58	57	67
68	68	67	66	65	64	63	62	62	61	60	59	58	68
69	69	68	67	66	65	64	63	63	62	61	60	59	69
70	70	69	68	67	67	66	65	64	63	62	61	61	70
71	71	70	69	68	68	67	66	65	64	63	62	62	71
72	72	71	70	69	69	68	67	66	65	64	63	63	72
73	73	72	71	70	70	69	68	67	66	66	65	64	73
74	74	73	72	71	71	70	69	68	67	67	66	65	74
75	75	74	73	72	72	71	70	69	68	68	67	66	75
76	76	75	74	73	73	72	71	70	69	69	68	67	76
77	77	76	75	74	74	73	72	71	70	70	69	68	77
78	78	77	76	75	75	74	73	72	71	71	70	69	78
79	79	78	77	76	76	75	74	73	72	72	71	70	79
80	80	79	78	77	77	76	75	74	73	73	72	72	80
t.	0°.5	1°.0	1°.5	2°.0	2°.5	3°.0	3°.5	4°.0	4°.5	5°.0	5°.5	6°.0	t.

TABLE I. — TEMPERATURE OF THE DEW-POINT, IN DEGREES FAHRENHEIT.

t (Dry ther.)	Difference between the dry and wet thermometers ($t-t'$).												t (Dry ther.)
	6°.0	6°.5	7°.0	7°.5	8°.0	8°.5	9°.0	9°.5	10°.0	10°.5	11°.0	11°.5	12°.0	
19	−25													19
20	−19	−32												20
21	−15	−24	−47											21
22	−11	−19	−31											22
23	− 8	−14	−24	−45										23
24	− 5	−10	−18	−30										24
25	− 2	− 7	−13	−22	−42									25
26	0	− 4	− 9	−17	−28									26
27	+ 3	− 1	− 6	−12	−20	−37								27
28	5	+ 1	− 3	− 8	−15	−25	−54							28
29	7	4	0	− 4	−10	−18	−32							29
30	9	6	+ 2	− 2	− 6	−13	−22	−43						30
31	11	8	5	+ 1	− 3	− 8	−15	−27						31
32	13	10	7	4	0	− 4	−10	−18	−33					32
33	14	12	9	6	+ 3	− 1	− 6	−12	−22	−44				33
34	16	14	11	8	6	+ 2	− 2	− 8	−15	−27				34
35	18	15	13	10	8	5	+ 1	− 4	− 9	−18	−32			35
36	19	17	15	12	10	8	4	0	− 5	−12	−20	−42		36
37	21	19	17	14	12	9	6	+ 3	− 2	− 6	−14	−25	−52	37
38	22	20	19	16	14	11	9	6	+ 2	− 2	− 8	−16	−29	38
39	24	22	20	18	16	14	11	8	5	+ 1	− 4	−10	−18	39
40	25	23	22	20	18	16	13	11	8	4	0	− 5	−12	40
41	26	25	23	21	20	17	15	13	10	7	+ 4	− 1	− 6	41
42	27	26	24	23	21	19	18	15	12	10	7	+ 3	− 2	42
43	29	27	26	24	23	21	19	17	14	12	9	6	+ 2	43
44	30	28	27	26	24	22	20	18	16	14	12	9	6	44
45	31	30	28	27	25	24	22	20	18	16	13	11	8	45
46	32	31	30	28	27	25	24	22	20	18	16	13	11	46
47	33	32	31	29	28	26	25	23	22	20	18	15	13	47
48	35	33	32	30	29	28	26	25	23	21	20	17	15	48
49	36	34	33	32	31	29	28	26	25	23	21	19	17	49
50	37	35	34	33	32	31	29	28	26	24	23	21	19	50
51	38	37	36	34	33	32	31	29	28	26	24	22	21	51
52	40	38	37	36	34	33	32	30	29	28	26	24	23	52
53	41	39	38	37	36	34	33	32	30	29	28	26	26	53
54	42	41	40	39	37	36	34	33	32	30	29	27	26	54
55	43	42	41	40	39	37	36	34	33	32	30	29	28	55
56	44	43	42	41	40	39	37	36	34	33	32	30	29	56
57	46	45	44	42	41	40	39	37	36	35	33	32	30	57
58	47	46	45	44	42	41	40	39	37	36	35	33	32	58
59	48	47	46	45	44	43	41	40	39	38	36	35	33	59
60	49	48	47	46	45	44	43	41	40	39	38	36	35	60
61	50	49	48	47	46	45	44	43	42	41	39	38	36	61
62	52	51	50	49	48	47	45	44	43	42	41	39	38	62
63	53	52	51	50	49	48	47	45	44	43	42	41	39	63
64	54	53	52	51	50	49	48	47	46	45	43	42	41	64
65	55	54	53	52	51	50	49	48	47	46	45	43	42	65
66	56	55	54	53	52	51	50	49	48	47	46	45	44	66
67	57	56	55	55	54	53	52	51	50	48	47	46	45	67
68	58	57	57	56	55	54	53	52	51	50	49	47	46	68
69	59	58	58	57	56	55	54	53	52	51	50	49	48	69
70	61	60	59	58	57	56	55	54	53	52	51	50	49	70
71	62	61	60	59	58	57	56	55	55	54	53	52	51	71
72	63	62	61	60	59	58	57	57	56	55	54	53	52	72
73	64	63	62	62	61	60	59	58	57	56	55	54	53	73
74	65	64	63	63	62	61	60	59	58	57	56	55	54	74
75	66	65	64	64	63	62	61	60	59	58	57	56	56	75
76	67	66	65	65	64	63	62	61	61	60	59	58	57	76
77	68	67	67	66	65	64	63	62	62	61	60	59	58	77
78	69	68	68	67	66	66	64	63	63	62	61	60	59	78
79	70	69	69	68	67	67	66	65	64	63	62	61	61	79
80	72	71	70	69	68	68	67	66	65	64	63	62	62	80
t.	6°.0	6°.5	7°.0	7°.5	8°.0	8°.5	9°.0	9°.5	10°.0	10°.5	11°.0	11°.5	12°.0	t.

150   OBSERVATIONAL METEOROLOGY.

TABLE I. — TEMPERATURE OF THE DEW-POINT, IN DEGREES FAHRENHEIT.

t (Dry ther.)	Difference between the dry and wet thermometers ($t-t'$).												t (Dry ther.)
	12°.0	12°.5	13°.0	13°.5	14°.0	14°.5	15°.0	15°.5	16°.0	16°.5	17°.0	17°.5	18°.0	
40	−12	−22	−44											40
41	− 6	−13	−25											41
42	− 2	− 7	−15	−28										42
43	+ 2	− 3	− 8	−17	−33									43
44	6	+ 1	− 4	−10	−19	−40								44
45	8	5	0	− 4	−11	−22	−48							45
46	11	8	+ 4	0	− 5	−13	−24							46
47	13	10	7	+ 3	− 1	− 6	−14	−27						47
48	15	12	10	6	+ 2	− 2	− 8	−16	−30					48
49	17	14	12	9	6	+ 2	− 3	− 9	−18	−35				49
50	19	16	14	12	9	5	+ 1	− 4	−10	−20	−42			50
51	21	18	17	14	11	8	5	0	− 5	−12	−22	−52		51
52	23	21	19	16	14	11	8	+ 4	0	− 6	−13	−25		52
53	24	22	20	18	16	14	11	8	+ 4	− 1	− 6	−14	−28	53
54	26	24	22	20	18	16	13	10	7	+ 3	− 2	− 8	−16	54
55	28	26	24	22	20	18	16	13	10	7	+ 3	− 2	− 8	55
56	29	27	26	24	22	20	18	15	13	10	6	+ 2	− 2	56
57	30	29	28	26	24	22	20	18	15	13	10	6	+ 2	57
58	32	30	29	27	26	24	22	20	18	15	12	9	6	58
59	33	32	31	29	27	26	24	22	20	18	15	12	9	59
60	35	33	32	30	29	27	26	24	22	20	18	15	12	60
61	36	35	33	32	31	29	28	26	24	22	20	18	15	61
62	38	37	35	34	32	31	29	28	26	24	22	20	18	62
63	39	38	37	35	34	32	31	29	28	26	24	22	20	63
64	41	39	38	37	35	34	32	31	29	28	26	24	22	64
65	42	41	40	38	37	35	34	32	31	29	28	26	24	65
66	44	43	41	40	38	37	35	34	32	31	30	28	26	66
67	45	44	43	41	40	39	37	36	34	32	31	30	28	67
68	46	45	44	43	42	40	39	38	36	34	33	31	30	68
69	48	47	46	45	43	42	40	39	38	36	34	33	32	69
70	49	48	47	46	45	43	42	41	39	38	36	35	33	70
71	51	49	48	47	46	45	43	42	41	39	38	36	35	71
72	52	51	50	49	47	46	45	44	43	41	40	38	37	72
73	53	52	51	50	49	48	46	45	44	43	41	40	38	73
74	54	53	52	51	50	49	48	47	45	44	43	41	40	74
75	56	55	54	53	52	50	49	48	47	45	44	43	42	75
76	57	56	55	54	53	52	50	49	48	47	46	45	43	76
77	58	57	56	55	54	53	52	51	50	49	48	46	45	77
78	59	58	57	56	55	54	53	52	51	50	49	48	47	78
79	61	60	59	58	57	56	55	54	53	52	51	49	48	79
80	62	61	60	59	58	57	56	55	54	53	52	51	50	80
t.	12°.0	12°.5	13°.0	13°.5	14°.0	14°.5	15°.0	15°.5	16°.0	16°.5	17°.0	17°.5	18°.0	t.

METEOROLOGICAL TABLES.   151

TABLE I. — TEMPERATURE OF THE DEW-POINT, IN DEGREES FAHRENHEIT.

t (Dry ther.)	Difference between the dry and wet thermometers ($t-t'$).													t (Dry ther.)
	18°.0	19°.0	20°.0	21°.0	22°.0	23°.0	24°.0	25°.0	26°.0	27°.0	28°.0	29°.0	30°.0	
55	−8													55
56	−2	−19												56
57	+2	−10	−48											57
58	6	− 3	−22											58
59	9	+ 1	−12											59
60	12	5	− 5	−25										60
61	15	9	0	−14										61
62	18	12	+ 5	− 6	−28									62
63	20	15	9	0	−14									63
64	22	18	12	+ 4	− 6	−32								64
65	24	20	15	9	0	−16								65
66	26	22	18	12	+ 4	− 7	−34							66
67	28	24	20	15	9	− 1	−16							67
68	30	26	23	18	12	+ 4	− 7	−37						68
69	32	28	25	20	15	8	0	−17						69
70	33	30	27	23	19	12	+ 5	− 7	−39					70
71	35	32	29	25	21	16	9	0	−17					71
72	37	33	31	27	23	18	13	+ 5	− 6	−30				72
73	38	35	32	29	25	21	16	10	0	−16				73
74	40	37	34	31	28	24	19	13	+ 6	− 6	−37			74
75	42	39	36	32	30	26	22	16	10	0	−16			75
76	43	41	38	34	32	28	24	20	14	+ 6	− 6	−34		76
77	45	42	40	36	33	30	26	22	17	11	+ 1	−14		77
78	47	44	41	38	35	32	28	24	20	14	7	− 4	−30	78
79	48	46	43	40	37	34	31	27	23	18	11	+ 2	−13	79
80	50	47	45	42	39	36	32	29	25	21	15	8	− 3	80
t.	18°.0	19°.0	20°.0	21°.0	22°.0	23°.0	24°.0	25°.0	26°.0	27°.0	28°.0	29°.0	30°.0	t.

OBSERVATIONAL METEOROLOGY.

TABLE I. — TEMPERATURE OF THE DEW-POINT, IN DEGREES FAHRENHEIT.

t (Dry ther.)	Difference between the dry and wet thermometers ($t-t'$).												t (Dry ther.)
	1°.0	2°.0	3°.0	4°.0	5°.0	6°.0	7°.0	8°.0	9°.0	10°.0	11°.0	12°.0	
80	79	77	76	74	73	72	70	68	67	65	63	62	80
81	80	78	77	75	74	73	71	70	68	66	65	63	81
82	81	79	78	77	75	74	72	71	69	68	66	64	82
83	82	80	79	78	76	75	73	72	70	69	67	65	83
84	83	81	80	79	77	76	74	73	71	70	68	67	84
85	84	82	81	80	78	77	75	74	72	71	69	68	85
86	85	83	82	81	79	78	76	75	73	72	71	69	86
87	86	84	83	82	80	79	78	76	74	73	72	70	87
88	87	85	84	83	81	80	79	77	75	74	73	71	88
89	88	86	85	84	82	81	80	78	76	76	74	72	89
90	89	87	86	85	84	82	81	79	78	77	75	74	90
91	90	88	87	86	85	83	82	80	79	78	76	75	91
92	91	89	88	87	86	84	83	82	80	79	77	76	92
93	92	91	89	88	87	85	84	83	81	80	78	77	93
94	93	92	90	89	88	86	85	84	82	81	80	78	94
95	94	93	91	90	89	87	86	85	83	82	81	79	95
96	95	94	92	91	90	88	87	86	84	83	82	80	96
97	96	95	93	92	91	90	88	87	86	84	83	81	97
98	97	96	94	93	92	91	89	88	87	85	84	83	98
99	98	97	95	94	93	92	90	89	88	86	85	84	99
100	99	98	96	95	94	93	91	90	89	87	86	85	100
101	100	99	97	96	95	94	92	91	90	88	87	86	101
102	101	100	98	97	96	95	93	92	91	90	88	87	102
103	102	101	99	98	97	96	94	93	92	91	89	88	103
104	103	102	100	99	98	97	96	94	93	92	90	89	104
105	104	103	101	100	99	98	97	95	94	93	91	90	105
106	105	104	102	101	100	99	98	96	95	94	93	91	106
107	106	105	103	102	101	100	99	97	96	95	94	92	107
108	107	106	104	103	102	101	100	98	97	96	95	93	108
109	108	107	105	104	103	102	101	99	98	97	96	94	109
110	109	108	107	105	104	103	102	101	99	98	97	96	110
t.	1°.0	2°.0	3°.0	4°.0	5°.0	6°.0	7°.0	8°.0	9°.0	10°.0	11°.0	12°.0	t.

TABLE I.—TEMPERATURE OF THE DEW-POINT, IN DEGREES FAHRENHEIT.

t (Dry ther.)	Difference between the dry and wet thermometers ($t-t'$).												t (Dry ther.)
	12°.0	13°.0	14°.0	15°.0	16°.0	17°.0	18°.0	19°.0	20°.0	21°.0	22°.0	23°.0	24°.0	
80	62	60	58	56	54	52	50	47	45	42	39	36	32	80
81	63	61	59	57	55	53	51	49	47	44	41	38	35	81
82	64	62	61	59	57	55	53	50	48	45	43	40	37	82
83	65	64	62	60	58	56	54	52	50	47	44	42	39	83
84	67	65	63	61	59	57	55	53	51	49	46	43	41	84
85	68	66	64	62	61	59	57	55	53	50	48	45	42	85
86	69	67	66	64	62	60	58	56	54	52	49	47	44	86
87	70	68	67	65	63	61	59	57	55	53	51	48	46	87
88	71	70	68	66	64	63	61	59	57	55	53	50	48	88
89	72	71	69	67	66	64	62	60	58	56	54	52	49	89
90	74	72	70	69	67	65	63	62	60	58	56	53	51	90
91	75	73	72	70	68	67	65	63	61	59	57	55	53	91
92	76	74	73	71	69	68	66	64	62	60	58	56	54	92
93	77	75	74	72	71	69	67	66	64	62	60	58	56	93
94	78	77	75	73	72	70	69	67	65	63	61	59	57	94
95	79	78	76	75	73	71	70	68	66	64	63	61	59	95
96	80	79	77	76	74	73	71	69	68	66	64	62	60	96
97	81	80	78	77	75	74	72	71	69	67	65	63	61	97
98	83	81	80	78	77	75	73	72	70	68	67	65	63	98
99	84	82	81	79	78	76	75	73	71	70	68	66	64	99
100	85	83	82	80	79	77	76	74	73	71	69	67	66	100
101	86	84	83	82	80	79	77	75	74	72	71	69	67	101
102	87	85	84	83	81	80	78	77	75	73	72	70	68	102
103	88	87	85	84	82	81	79	78	76	75	73	71	70	103
104	89	88	86	85	83	82	81	79	78	76	74	73	71	104
105	90	89	87	86	85	83	82	80	79	77	76	74	72	105
106	91	90	89	87	86	84	83	81	80	78	77	75	74	106
107	92	91	90	88	87	85	84	83	81	80	78	76	75	107
108	93	92	91	89	88	87	85	84	82	81	79	78	76	108
109	94	93	92	90	89	88	86	85	83	82	80	79	77	109
110	96	94	93	92	90	89	87	86	85	83	82	80	79	110
t.	12°.0	13°.0	14°.0	15°.0	16°.0	17°.0	18°.0	19°.0	20°.0	21°.0	22°.0	23°.0	24°.0	t.

TABLE I. — TEMPERATURE OF THE DEW-POINT, IN DEGREES FAHRENHEIT.

(Dry ther.)	Difference between the dry and wet thermometers (t—t').													(Dry ther.)
	24°.0	25°.0	26°.0	27°.0	28°.0	29°.0	30°.0	31°.0	32°.0	33°.0	34°.0	35°.0	36°.0	
80	32	29	25	21	15	8	− 3	−27						80
81	35	31	28	24	18	12	+ 3	−11						81
82	37	33	30	26	22	16	9	− 2	−24					82
83	39	35	32	28	24	19	13	+ 5	− 9					83
84	41	37	34	30	27	22	17	10	0	−20				84
85	42	39	36	32	29	25	20	14	+ 6	− 7	−54			85
86	44	41	38	35	31	28	23	18	11	+ 1	−17			86
87	46	43	40	37	33	30	26	21	15	7	− 5	−38		87
88	48	45	42	39	35	32	28	24	19	12	+ 3	−13		88
89	49	47	44	41	38	34	31	27	22	16	9	− 2	−28	89
90	51	48	46	43	40	36	32	29	25	20	13	+ 4	−10	90
91	53	50	47	45	42	38	35	32	28	23	18	10	0	91
92	54	52	49	46	44	41	37	34	30	26	21	15	+ 7	92
93	56	53	51	48	46	43	39	36	32	29	24	19	12	93
94	57	55	53	50	47	45	42	38	35	31	27	22	16	94
95	59	56	54	52	49	46	44	40	37	33	30	25	20	95
96	60	58	56	53	51	48	46	43	39	36	32	28	24	96
97	61	59	57	55	53	50	47	45	41	38	34	31	26	97
98	63	61	59	57	54	52	49	47	44	40	37	33	29	98
99	64	62	60	58	56	54	51	48	46	43	39	35	32	99
100	66	64	62	60	57	55	53	50	48	45	41	38	34	100
101	67	65	63	61	59	57	54	52	49	47	44	40	37	101
102	68	66	65	63	61	58	56	54	51	49	46	43	39	102
103	70	68	66	64	62	60	58	55	53	50	48	45	41	103
104	71	69	67	65	63	61	59	57	55	52	50	47	44	104
105	72	70	69	67	65	63	61	59	56	54	52	49	46	105
106	74	72	70	68	66	64	62	60	58	56	53	51	48	106
107	75	73	71	70	68	66	64	62	60	57	55	52	50	107
108	76	74	73	71	69	67	65	63	61	59	57	54	52	108
109	77	76	74	72	71	69	67	65	63	61	58	56	54	109
110	79	77	75	74	72	70	68	66	64	62	60	58	55	110
t.	24°.0	25°.0	26°.0	27°.0	28°.0	29°.0	30°.0	31°.0	32°.0	33°.0	34°.0	35°.0	36°.0	t.

METEOROLOGICAL TABLES. 155

TABLE I. — TEMPERATURE OF THE DEW-POINT, IN DEGREES FAHRENHEIT.

t (Dry ther.)	Difference between the dry and wet thermometers ($t-t'$).												t (Dry ther.)
	36°.0	37°.0	38°.0	39°.0	40°.0	41°.0	42°.0	43°.0	44°.0	45°.0	46°.0	47°.0	48°.0	
89	−28													89
90	−10													90
91	0	−22												91
92	+ 7	− 7	−16											92
93	12	+ 2	− 4											93
94	16	8	+ 4	−37										94
95	20	13	10	−12										95
96	24	15	15	− 1	−25									96
97	26	21	19	+ 7	− 8	−18								97
98	29	25	23	12	+ 2	− 4								98
99	32	28	26	17	9		−42							99
100	34	30	29	21	14	+ 5	−12							100
101	37	32	32	24	18	11	0	−25						101
102	39	35	34	27	22	16	+ 7	− 7						102
103	41	38	34	30	26	20	13	+ 3	−16					103
104	44	40	37	32	29	24	18	10	− 2	−38				104
105	46	43	39	35	31	27	22	15	+ 6	−10				105
106	48	45	42	38	34	30	25	20	12	+ 1	−22			106
107	50	47	44	40	37	32	28	24	17	9	− 5			107
108	52	49	46	43	39	35	31	27	21	14	+ 5	−13		108
109	54	51	48	45	42	38	34	30	25	19	12	0	−28	109
110	55	53	50	47	44	41	37	33	28	23	17	+ 8	− 7	110
t.	36°.0	37°.0	38°.0	39°.0	40°.0	41°.0	42°.0	43°.0	44°.0	45°.0	46°.0	47°.0	48°.0	t.

Table II. — Relative Humidity, per cent.

t (Dry ther.)	\multicolumn{12}{c	}{Difference between the dry and wet thermometers (t − t′).}	t (Dry ther.)
	0°.2	0°.4	0°.6	0°.8	1°.0	1°.2	1°.4	1°.6	1°.8	2°.0	2°.2	2°.4	2°.6	
−40	46													−40
−39	49													−39
−38	51													−38
−37	54													−37
−36	56													−36
−35	59													−35
−34	61	22												−34
−33	63	25												−33
−32	65	30												−32
−31	67	34												−31
−30	69	38												−30
−29	71	42												−29
−28	72	45	17											−28
−27	74	48	22											−27
−26	76	51	26											−26
−25	77	53	31											−25
−24	78	56	34	12										−24
−23	79	58	37	16										−23
−22	80	60	40	20										−22
−21	81	62	44	25										−21
−20	82	64	47	29	11									−20
−19	83	66	49	33	16									−19
−18	84	68	52	36	20									−18
−17	85	70	54	39	24	9								−17
−16	86	71	57	43	28	14								−16
−15	86	73	59	46	32	19								−15
−14	87	74	61	48	36	23	10							−14
−13	88	76	63	51	39	27	15							−13
−12	88	77	65	53	42	30	19	7						−12
−11	89	78	67	56	45	34	23	12						−11
−10	90	79	68	58	48	37	26	16	5					−10
−9	90	80	70	60	50	40	30	20	10					−9
−8	90	81	71	62	52	43	33	24	14	5				−8
−7	91	82	73	63	54	45	36	27	18	9				−7
−6	91	83	74	65	56	48	39	31	22	13	5			−6
−5	92	83	75	67	58	50	42	34	25	17	9			−5
−4	92	84	76	68	60	52	45	37	29	21	13	5		−4
−3	92	85	77	70	62	55	47	40	32	25	17	10		−3
−2	93	86	78	71	64	57	50	42	35	28	21	14	7	−2
−1	93	86	79	72	66	59	52	45	38	31	25	18	11	−1
0	93	87	80	74	67	61	54	48	41	35	28	22	15	0
+1	94	87	81	75	69	63	56	50	44	38	32	25	19	+1
2	94	88	82	76	70	64	58	52	46	40	35	29	23	2
3	94	88	83	77	71	66	60	54	49	43	37	32	26	3
4	94	89	83	78	73	67	62	56	51	45	40	34	29	4
5	95	89	84	79	74	68	63	58	53	48	42	37	32	5
6	95	90	85	80	75	70	65	60	54	50	44	39	34	6
7	95	90	85	80	76	71	66	61	56	51	47	42	37	7
8	95	91	86	81	76	72	67	62	58	53	49	44	39	8
9	96	91	86	82	77	73	68	64	59	55	51	46	42	9
10	96	91	87	83	78	74	69	65	61	57	52	48	44	10
11	96	92	87	83	79	75	71	66	62	58	54	50	46	11
12	96	92	88	84	80	76	72	68	64	60	56	52	48	12
13	96	92	88	84	81	77	73	69	65	61	58	54	50	13
14	96	93	89	85	81	78	74	70	67	63	59	56	52	14
15	96	93	89	86	82	79	75	71	68	64	61	57	54	15
16	97	93	90	86	83	79	76	73	69	66	62	59	56	16
17	97	93	90	87	83	80	77	74	70	67	64	60	57	17
18	97	94	90	87	84	81	78	74	71	68	65	62	59	18
19	97	94	91	88	84	81	78	75	72	69	66	63	60	19
20	97	94	91	88	85	82	79	76	73	70	67	64	61	20
t.	0°.2	0°.4	0°.6	0°.8	1°.0	1°.2	1°.4	1°.6	1°.8	2°.0	2°.2	2°.4	2°.6	t.

METEOROLOGICAL TABLES.

TABLE II. — RELATIVE HUMIDITY, PER CENT.

t (Dry ther.)	Difference between the dry and wet thermometers ($t - t'$).										t (Dry ther.)
	2°.6	2°.8	3°.0	3°.2	3°.4	3°.6	3°.8	4°.0	4°.2	4°.4	
− 2	7										− 2
− 1	11	4									− 1
0	15	9									0
+ 1	19	13	7								+ 1
2	23	17	11	5							2
3	26	20	15	9	4						3
4	29	24	18	13	8	2					4
5	32	27	22	16	11	6					5
6	34	29	25	20	15	10	5				6
7	37	32	28	23	18	13	9	4			7
8	39	35	30	26	21	17	12	8	3		8
9	42	37	33	28	24	20	15	11	7	2	9
10	44	40	35	31	27	23	19	14	10	6	10
11	46	42	38	34	30	26	22	18	14	10	11
12	48	44	40	36	32	28	25	21	17	13	12
13	50	46	42	39	35	31	27	24	20	16	13
14	52	48	45	41	37	34	30	27	23	19	14
15	54	50	47	43	40	36	33	29	26	23	15
16	56	52	49	46	42	39	36	32	29	25	16
17	57	54	51	48	44	41	38	35	31	28	17
18	59	56	53	49	46	43	40	37	34	31	18
19	60	57	54	51	48	45	42	39	36	33	19
20	61	58	56	53	50	47	44	41	38	35	20
t.	2°.6	2°.8	3°.0	3°.2	3°.4	3°.6	3°.8	4°.0	4°.2	4°.4	t.

t.	4°.6	4°.8	5°.0	5°.2	5°.4	5°.6	5°.8	6°.0	6°.2	6°.4	t.
8											8
9											9
10	2										10
11	6	2									11
12	9	5	2								12
13	13	9	5	2							13
14	16	12	9	5	.2						14
15	19	16	12	9	5	2					15
16	22	19	16	12	9	6	2				16
17	25	22	19	16	12	9	6	3			17
18	28	25	22	19	16	13	9	6	3		18
19	30	27	24	21	19	16	13	10	7	4	19
20	33	30	27	24	21	19	16	13	10	7	20
t.	4°.6	4°.8	5°.0	5°.2	5°.4	5°.6	5°.8	6°.0	6°.2	6°.4	t.

TABLE II. — Relative Humidity, per cent.

t (Dry ther.)	\multicolumn{11}{c}{Difference between the dry and wet thermometers ($t-t'$).}	t (Dry ther.)
	0°.5	1°.0	1°.5	2°.0	2°.5	3°.0	3°.5	4°.0	4°.5	5°.0	5°.5	6°.0	
20	92	85	77	70	63	56	48	41	34	27	20	13	20
21	93	85	78	71	64	57	50	43	36	29	23	16	21
22	93	86	79	72	65	58	51	45	38	32	25	19	22
23	93	86	80	73	66	60	53	46	40	34	27	21	23
24	93	87	80	74	67	61	54	48	42	36	30	24	24
25	94	87	81	74	68	62	56	50	44	38	32	26	25
26	94	88	81	75	69	63	57	51	45	40	34	28	26
27	94	88	82	76	70	64	59	53	47	42	36	30	27
28	94	88	82	77	71	65	60	54	49	43	38	33	28
29	94	89	83	77	72	66	61	56	50	45	40	35	29
30	94	89	84	78	73	67	62	57	52	47	41	36	30
31	95	89	84	79	74	68	63	58	53	48	43	38	31
32	95	90	84	79	74	69	64	59	54	50	45	40	32
33	95	90	85	80	75	70	65	60	56	51	47	42	33
34	95	91	86	81	75	72	67	62	57	53	48	44	34
35	95	91	86	82	76	73	69	65	59	54	50	45	35
36	96	91	86	82	77	73	70	66	61	56	51	47	36
37	96	91	87	82	78	74	70	66	62	57	52	48	37
38	96	92	87	83	79	75	71	67	63	58	54	50	38
39	96	92	88	83	79	75	72	68	63	59	55	52	39
40	96	92	88	84	80	76	72	68	64	60	56	53	40
41	96	92	88	84	80	76	72	69	65	61	57	54	41
42	96	92	88	84	81	77	73	69	65	62	58	55	42
43	96	92	88	85	81	77	74	70	66	63	59	56	43
44	96	92	88	85	81	78	74	70	67	63	60	57	44
45	96	92	89	85	82	78	75	71	67	64	61	58	45
46	96	93	89	85	82	79	75	72	68	65	61	58	46
47	96	93	89	86	83	79	76	72	69	66	62	59	47
48	96	93	89	86	83	79	76	73	69	66	63	60	48
49	97	93	90	86	83	80	76	73	70	67	63	60	49
50	97	93	90	87	83	80	77	74	70	67	64	61	50
51	97	93	90	87	84	81	77	74	71	68	65	62	51
52	97	94	90	87	84	81	78	75	72	69	66	63	52
53	97	94	91	87	84	81	78	75	72	69	66	63	53
54	97	94	91	88	85	82	79	76	73	70	67	64	54
55	97	94	91	88	85	82	79	76	73	70	68	65	55
56	97	94	91	88	85	82	80	77	74	71	69	65	56
57	97	94	91	88	86	83	80	77	74	71	69	66	57
58	97	94	91	89	86	83	80	78	75	72	69	67	58
59	97	94	92	89	86	83	81	78	75	72	70	67	59
60	97	94	92	89	86	84	81	78	75	73	70	68	60
61	97	94	92	89	87	84	81	78	76	73	71	68	61
62	97	95	92	89	87	84	81	79	76	74	71	69	62
63	97	95	92	89	87	84	82	79	77	74	72	69	63
64	97	95	92	90	87	85	82	79	77	74	72	70	64
65	97	95	92	90	87	85	82	80	77	75	72	70	65
66	97	95	92	90	87	85	82	80	78	75	73	71	66
67	98	95	93	90	88	85	83	80	78	76	73	71	67
68	98	95	93	90	88	85	83	81	78	76	74	71	68
69	98	95	93	90	88	86	83	81	78	76	74	72	69
70	98	95	93	90	88	86	83	81	79	77	74	72	70
71	98	95	93	91	88	86	84	81	79	77	75	72	71
72	98	95	93	91	88	86	84	82	79	77	75	73	72
73	98	95	93	91	88	86	84	82	80	78	75	73	73
74	98	95	93	91	88	86	84	82	80	78	76	74	74
75	98	95	93	91	89	87	84	82	80	78	76	74	75
76	98	95	93	91	89	87	85	82	80	78	76	74	76
77	98	95	93	91	89	87	85	83	80	78	76	74	77
78	98	96	93	91	89	87	85	83	81	79	77	75	78
79	98	96	94	91	89	87	85	83	81	79	77	75	79
80	98	96	94	92	89	87	85	83	81	79	77	75	80
t	0°.5	1°.0	1°.5	2°.0	2°.5	3°.0	3°.5	4°.0	4°.5	5°.0	5°.5	6°.0	t

TABLE II. — RELATIVE HUMIDITY, PER CENT.

(Dry ther.)	Difference between the dry and wet thermometers ($t-t'$).												(Dry ther.)
	6°.0	6°.5	7°.0	7°.5	8°.0	8°.5	9°.0	9°.5	10°.0	10°.5	11°.0	11°.5	12°.0	
19	10													19
20	13	6												20
21	16	9	2											21
22	19	12	6											22
23	21	15	9	2										23
24	24	17	11	6										24
25	26	20	14	8	3									25
26	28	23	17	11	6									26
27	30	25	19	14	9	3								27
28	33	27	22	17	11	6	1							28
29	35	29	24	19	14	9	4							29
30	36	31	26	22	17	12	7	2						30
31	38	33	29	24	19	14	10	5						31
32	40	35	31	26	21	17	12	8	3					32
33	42	37	33	28	24	19	15	10	6	2				33
34	44	39	35	30	26	21	17	13	9	4				34
35	45	41	37	32	28	24	19	15	12	7	3			35
36	47	43	38	34	30	26	22	18	14	10	6	2		36
37	48	44	40	36	32	28	24	20	16	12	8	5	1	37
38	50	46	42	38	34	30	26	22	18	15	11	7	3	38
39	52	48	44	40	36	32	28	24	20	17	13	9	6	39
40	53	49	45	41	38	34	30	26	22	19	16	12	8	40
41	54	50	46	43	39	36	32	29	24	21	18	14	10	41
42	55	51	48	44	40	37	34	30	27	23	20	16	13	42
43	56	52	49	46	42	38	35	32	29	25	22	19	15	43
44	57	53	50	47	43	40	37	33	30	27	24	21	17	44
45	58	54	51	48	44	41	38	35	32	29	25	22	19	45
46	58	55	52	49	46	42	39	36	33	30	27	23	21	46
47	59	56	53	50	47	44	40	38	34	31	28	25	22	47
48	60	56	53	51	48	45	42	39	36	33	30	27	24	48
49	60	57	54	52	49	46	43	40	37	34	31	29	26	49
50	61	58	55	52	50	47	44	41	38	36	33	30	27	50
51	62	59	56	53	50	48	45	42	39	37	34	31	28	51
52	63	60	57	54	51	48	46	43	40	38	35	33	30	52
53	63	61	58	55	52	49	47	44	42	39	36	34	31	53
54	64	61	59	56	53	50	48	45	43	40	38	35	32	54
55	65	62	59	57	54	51	49	46	41	39	36	34	55
56	65	63	60	57	55	52	50	47	44	42	40	37	35	56
57	66	64	61	58	55	53	50	48	45	43	40	38	36	57
58	67	64	61	59	56	53	51	49	46	44	42	39	37	58
59	67	65	62	60	57	54	52	49	47	45	43	40	38	59
60	68	65	63	60	58	55	53	50	48	46	44	41	39	60
61	68	66	63	61	58	56	54	51	49	47	44	42	40	61
62	69	66	64	61	59	57	54	52	50	47	45	43	41	62
63	69	67	64	62	60	57	55	53	51	48	46	44	42	63
64	70	67	65	62	60	58	56	53	51	49	47	45	43	64
65	70	68	65	63	61	59	56	54	52	50	48	46	44	65
66	71	68	66	63	61	59	57	55	53	51	49	47	45	66
67	71	69	66	64	62	60	58	55	53	51	49	47	45	67
68	71	69	67	65	63	60	58	56	54	52	50	48	46	68
69	72	70	67	65	63	61	59	57	55	53	51	49	47	69
70	72	70	68	66	64	62	60	57	55	53	52	50	48	70
71	72	70	68	66	64	62	60	58	56	54	52	50	48	71
72	73	71	69	67	65	63	61	59	57	55	53	51	49	72
73	73	71	69	67	65	63	61	59	57	55	53	52	50	73
74	74	72	70	68	66	64	62	60	58	56	54	52	50	74
75	74	72	70	68	66	64	62	60	58	56	55	53	51	75
76	74	72	70	68	66	64	63	61	59	57	55	53	52	76
77	74	73	71	69	67	65	63	61	59	57	56	54	52	77
78	75	73	71	69	67	65	63	62	60	58	56	54	53	78
79	75	73	71	70	68	66	64	62	60	58	57	55	53	79
80	75	73	72	70	68	66	64	63	61	59	57	55	54	80
$t.$	6°.0	6°.5	7°.0	7°.5	8°.0	8°.5	9°.0	9°.5	10°.0	10°.5	11°.0	11°.5	12°.0	$t.$

TABLE II. — RELATIVE HUMIDITY, PER CENT.

t (Dry ther.)	Difference between the dry and wet thermometers ($t-t'$).													t (Dry ther.)
	12°.0	12°.5	13°.0	13°.5	14°.0	14°.5	15°.0	15°.5	16°.0	16°.5	17°.0	17°.5	18°.0	
40	8	5	1											40
41	10	7	4											41
42	13	10	6	3										42
43	15	12	9	5	2									43
44	17	14	11	8	5	1								44
45	19	16	13	10	7	4	1							45
46	21	18	15	12	9	6	3							46
47	22	20	16	14	11	8	5	3						47
48	24	21	19	16	13	10	7	5	2					48
49	26	23	20	17	15	12	9	7	4	1				49
50	27	24	22	19	16	14	11	9	6	4	1			50
51	28	26	23	21	18	16	13	10	8	5	3			51
52	30	27	24	22	20	17	15	12	10	7	5			52
53	31	29	26	24	21	19	16	14	12	9	7	4	2	53
54	32	30	28	25	23	20	18	15	13	11	8	6	4	54
55	34	31	29	26	24	22	19	17	15	12	10	8	6	55
56	35	33	30	28	25	23	21	19	16	14	12	10	8	56
57	36	34	32	29	27	24	22	20	18	16	13	11	9	57
58	37	35	33	30	28	26	24	21	19	17	15	13	11	58
59	38	36	34	31	29	27	25	23	21	18	16	14	12	59
60	39	37	34	32	30	28	26	24	22	20	18	16	14	60
61	40	38	35	33	32	29	27	25	23	21	19	17	15	61
62	41	39	37	34	32	30	28	26	24	22	20	18	16	62
63	42	40	38	35	33	31	29	28	26	24	22	20	18	63
64	43	41	38	36	34	32	30	29	27	25	23	21	19	64
65	44	42	39	37	35	33	31	29	28	26	24	22	20	65
66	45	42	40	38	36	34	32	30	29	27	25	23	22	66
67	45	43	41	39	37	35	33	32	30	28	26	25	23	67
68	46	44	42	40	38	36	34	33	31	29	27	26	24	68
69	47	45	43	41	39	37	35	33	32	30	28	26	25	69
70	48	46	44	42	40	38	36	34	33	31	29	27	26	70
71	48	46	45	43	41	39	37	35	34	32	30	28	27	71
72	49	47	45	43	42	40	38	36	35	33	31	30	28	72
73	50	48	46	44	42	41	39	37	35	34	32	30	29	73
74	50	48	47	45	43	41	40	38	36	35	33	31	30	74
75	51	49	47	46	44	42	40	39	37	35	34	32	31	75
76	52	50	48	46	45	43	41	39	38	36	35	33	31	76
77	52	50	49	47	45	44	42	40	39	37	35	34	32	77
78	53	51	49	48	46	44	43	41	39	38	36	35	33	78
79	53	52	50	48	47	45	43	42	40	39	37	36	34	79
80	54	52	51	49	47	45	44	42	41	39	38	36	35	80
t	12°.0	12°.5	13°.0	13°.5	14°.0	14°.5	15°.0	15°.5	16°.0	16°.5	17°.0	17°.5	18°.0	t

TABLE II. — RELATIVE HUMIDITY, PER CENT.

t (Dry ther.)	Difference between the dry and wet thermometers ($t-t'$).													t (Dry ther.)
	18°.0	19°.0	20°.0	21°.0	22°.0	23°.0	24°.0	25°.0	26°.0	27°.0	28°.0	29°.0	30°.0	
55	6	1												55
56	8	3												56
57	9	5												57
58	11	7	2											58
59	12	8	4											59
60	14	10	6	2										60
61	15	11	7	3										61
62	16	13	9	5	1									62
63	18	14	10	7	3									63
64	19	15	12	8	5	1								64
65	20	17	13	10	6	3								65
66	22	18	14	11	8	4	1							66
67	23	19	16	12	9	6	2							67
68	24	20	17	14	10	7	4	1						68
69	25	22	18	15	12	8	5	2						69
70	26	23	19	16	13	10	7	4	1					70
71	27	24	20	17	14	11	8	5	2					71
72	28	24	22	18	15	12	9	6	3	1				72
73	29	25	22	19	16	13	10	8	5	2				73
74	30	26	23	20	18	15	12	9	6	3	1			74
75	31	27	24	21	19	16	13	10	7	5	2			75
76	31	28	25	22	20	17	14	11	8	6	3	1		76
77	32	29	26	23	20	18	15	12	10	7	4	2		77
78	33	30	27	24	21	19	16	13	11	8	6	3	1	78
79	34	31	28	25	22	19	17	14	12	9	7	4	2	79
80	35	32	29	26	23	20	18	15	13	10	8	6	3	80
t.	18°.0	19°.0	20°.0	21°.0	22°.0	23°.0	24°.0	25°.0	26°.0	27°.0	28°.0	29°.0	30°.0	t.

Table II. — Relative Humidity, per cent.

t (Dry ther.)	Difference between the dry and wet thermometers ($t-t'$).												t (Dry ther.)
	1°.0	2°.0	3°.0	4°.0	5°.0	6°.0	7°.0	8°.0	9°.0	10°.0	11°.0	12°.0	
80	96	92	87	83	79	75	72	68	64	61	57	54	80
81	96	92	88	84	80	76	72	68	65	61	58	54	81
82	96	92	88	84	80	76	72	69	65	62	58	55	82
83	96	92	88	84	80	76	73	69	66	62	59	55	83
84	96	92	88	84	80	77	73	69	66	63	59	56	84
85	96	92	88	84	80	77	73	70	66	63	60	56	85
86	96	92	88	84	81	77	73	70	67	63	60	57	86
87	96	92	88	84	81	77	74	70	67	64	60	57	87
88	96	92	88	85	81	77	74	71	67	64	61	58	88
89	96	92	88	85	81	78	74	71	68	64	61	58	89
90	96	92	88	85	81	78	75	71	68	65	62	59	90
91	96	92	89	85	82	78	75	71	68	65	62	59	91
92	96	92	89	85	82	78	75	72	69	65	62	59	92
93	96	93	89	85	82	78	75	72	69	66	63	60	93
94	96	93	89	86	82	79	75	72	69	66	63	60	94
95	96	93	89	86	82	79	76	72	69	66	63	60	95
96	96	93	89	86	82	79	76	73	70	67	64	61	96
97	96	93	89	86	82	79	76	73	70	67	64	61	97
98	96	93	89	86	83	79	76	73	70	67	64	61	98
99	96	93	89	86	83	80	76	73	70	68	65	62	99
100	97	93	90	86	83	80	77	74	71	68	65	62	100
101	97	93	90	86	83	80	77	74	71	68	65	62	101
102	97	93	90	86	83	80	77	74	71	68	65	63	102
103	97	93	90	87	83	80	77	74	71	69	66	63	103
104	97	93	90	87	83	80	77	74	72	69	66	63	104
105	97	93	90	87	84	81	78	75	72	69	66	64	105
106	97	93	90	87	84	81	78	75	72	69	66	64	106
107	97	93	90	87	84	81	78	75	72	69	67	64	107
108	97	93	90	87	84	81	78	75	72	70	67	64	108
109	97	93	90	87	84	81	78	75	73	70	67	65	109
110	97	94	90	87	84	81	78	76	73	70	67	65	110
t.	1°.0	2°.0	3°.0	4°.0	5°.0	6°.0	7°.0	8°.0	9°.0	10°.0	11°.0	12°.0	t.

TABLE II. — RELATIVE HUMIDITY, PER CENT.

t (Dry ther.)	Difference between the dry and wet thermometers ($t-t'$).													t (Dry ther.)
	12°.0	13°.0	14°.0	15°.0	16°.0	17°.0	18°.0	19°.0	20°.0	21°.0	22°.0	23°.0	24°.0	
80	54	51	47	44	41	38	35	32	29	26	23	20	18	80
81	54	51	48	44	41	38	35	33	30	27	24	21	19	81
82	55	52	48	45	42	39	36	33	31	28	25	22	20	82
83	55	52	49	46	43	40	37	34	31	29	26	23	21	83
84	56	53	49	46	44	41	38	35	32	29	27	24	22	84
85	56	53	50	47	44	41	38	36	33	30	28	25	22	85
86	57	54	51	48	45	42	39	36	34	31	29	26	23	86
87	57	54	51	48	45	42	40	37	34	32	30	27	24	87
88	58	55	52	49	46	43	40	38	35	32	30	27	25	88
89	58	55	52	49	46	44	41	38	36	33	31	28	26	89
90	59	56	53	50	47	44	41	39	36	34	32	29	26	90
91	59	56	53	50	47	45	42	39	37	35	33	30	27	91
92	59	56	54	51	48	45	43	40	37	35	33	30	28	92
93	60	57	54	51	48	46	43	41	38	36	34	31	29	93
94	60	57	54	52	49	46	44	41	39	36	34	31	29	94
95	60	58	55	52	49	47	44	42	39	37	35	32	30	95
96	61	58	55	53	50	47	45	42	40	37	36	33	30	96
97	61	58	56	53	50	48	45	43	40	38	36	33	31	97
98	61	59	56	53	51	48	46	43	41	38	37	34	32	98
99	62	59	56	54	51	49	46	44	41	39	37	34	32	99
100	62	59	57	54	51	49	47	44	42	39	37	35	33	100
101	62	60	57	54	52	49	47	45	42	40	38	36	33	101
102	63	60	57	55	52	50	47	45	43	40	38	36	34	102
103	63	60	58	55	53	50	48	45	43	41	39	37	34	103
104	63	61	58	55	53	51	48	46	44	41	39	37	35	104
105	64	61	58	56	53	51	49	46	44	42	40	38	35	105
106	64	61	59	56	54	51	49	47	44	42	40	38	36	106
107	64	62	59	57	54	52	49	47	45	43	41	38	36	107
108	64	62	59	57	54	52	50	47	45	43	41	39	37	108
109	65	62	60	57	55	52	50	48	46	44	41	39	37	109
110	65	62	60	57	55	53	50	48	46	44	42	40	38	110
t	12°.0	13°.0	14°.0	15°.0	16°.0	17°.0	18°.0	19°.0	20°.0	21°.0	22°.0	23°.0	24°.0	t

Table II. — Relative Humidity, per cent.

t (Dry ther.)	Difference between the dry and wet thermometers ($t-t'$).													t (Dry ther.)
	24°.0	25°.0	26°.0	27°.0	28°.0	29°.0	30°.0	31°.0	32°.0	33°.0	34°.0	35°.0	36°.0	
80	18	15	13	10	8	6	3	1						80
81	19	16	14	11	9	7	4	2						81
82	20	17	15	12	10	8	5	3	1					82
83	21	18	16	13	11	9	6	4	2					83
84	22	19	17	14	12	10	8	5	3	1				84
85	22	20	17	15	13	11	9	6	4	2				85
86	23	21	18	16	14	12	10	7	5	3	1			86
87	24	22	19	17	15	13	11	8	6	4	2			87
88	25	22	20	18	16	14	12	9	7	5	3	1		88
89	26	23	21	19	16	14	12	10	8	6	4	2	1	89
90	26	24	22	20	17	15	13	11	9	7	5	3	2	90
91	27	25	23	20	18	16	14	12	10	8	6	4	3	91
92	28	26	23	21	19	17	15	13	11	9	7	5	3	92
93	29	26	24	22	20	18	16	14	12	10	8	6	4	93
94	29	27	25	23	21	18	16	14	13	11	9	7	5	94
95	30	28	25	23	21	19	17	15	13	11	10	8	6	95
96	30	28	26	24	22	20	18	16	14	12	10	9	7	96
97	31	29	27	25	23	21	19	17	15	13	11	10	8	97
98	32	29	27	25	23	21	19	18	16	14	12	10	9	98
99	32	30	28	26	24	22	20	18	16	15	13	11	10	99
100	33	31	29	27	25	23	21	19	17	15	14	12	10	100
101	33	31	29	27	25	23	21	20	18	16	14	13	11	101
102	34	32	30	28	26	24	22	20	19	17	15	13	12	102
103	34	32	30	28	26	25	23	21	19	17	16	14	12	103
104	35	33	31	29	27	25	23	22	20	18	16	15	13	104
105	35	33	31	30	28	26	24	22	20	19	17	15	14	105
106	36	34	32	30	28	26	25	23	21	19	18	16	14	106
107	36	34	32	31	29	27	25	23	22	20	18	17	15	107
108	37	35	33	31	29	27	26	24	22	21	19	17	16	108
109	37	35	33	32	30	28	26	25	23	21	20	18	16	109
110	38	36	34	32	30	28	27	25	23	22	20	19	17	110
t.	24°.0	25°.0	26°.0	27°.0	28°.0	29°.0	30°.0	31°.0	32°.0	33°.0	34°.0	35°.0	36°.0	t.

TABLE II. — RELATIVE HUMIDITY, PER CENT.

t (Dry ther.)	Difference between the dry and wet thermometers (t—t').												t (Dry ther.)
	36°.0	37°.0	38°.0	39°.0	40°.0	41°.0	42°.0	43°.0	44°.0	45°.0	46°.0	47°.0	48°.0	
89	1													89
90	2													90
91	3	1												91
92	3	2	1											92
93	4	3	2											93
94	5	4	3	0										94
95	6	5	4	1										95
96	7	5	5	2	1									96
97	8	6	6	3	1									97
98	9	7	7	4	2	1								98
99	10	8	7	5	3	2	0							99
100	10	9	8	6	4	3	1							100
101	11	9	9	6	5	3	2	0						101
102	12	10	9	7	6	4	3	1						102
103	12	11	9	8	6	5	4	2	1					103
104	13	12	10	8	7	6	5	3	2					104
105	14	12	11	9	8	6	5	4	2	1				105
106	14	13	11	10	8	7	6	4	3	2	0			106
107	15	14	12	11	9	8	6	5	4	3	1			107
108	16	14	13	11	10	9	7	6	5	3	2	1		108
109	16	15	13	12	10	9	8	7	5	4	3	1		109
110	17	15	14	13	11	10	8	7	6	5	3	2	1	110
t.	36°.0	37°.0	38°.0	39°.0	40°.0	41°.0	42°.0	43°.0	44°.0	45°.0	46°.0	47°.0	48°.0	t.

TABLE III. REDUCTION OF BAROMETER READING TO 32°.

Temperature.	Inches.														
	24.0	24.5	25.0	25.5	26.0	26.5	27.0	27.5	28.0	28.5	29.0	29.5	30.0	30.5	31.0
30	−.003	−.003	−.003	−.003	−.003	−.003	−.003	−.003	−.003	−.004	−.004	−.004	−.004	−.004	−.004
31	.005	.005	.005	.005	.006	.006	.006	.006	.006	.006	.006	.006	.006	.007	.007
32	.007	.008	.008	.008	.008	.008	.008	.008	.009	.009	.009	.009	.009	.009	.009
33	.010	.010	.010	.010	.010	.010	.011	.011	.011	.011	.012	.012	.012	.012	.012
34	.012	.012	.012	.012	.013	.013	.013	.013	.014	.014	.014	.014	.015	.015	.015
35	.014	.014	.014	.015	.015	.015	.016	.016	.016	.016	.017	.017	.017	.018	.018
36	.016	.016	.017	.017	.017	.018	.018	.018	.019	.019	.019	.019	.020	.020	.021
37	.018	.019	.019	.019	.020	.020	.021	.021	.021	.022	.022	.022	.023	.023	.024
38	.020	.021	.021	.022	.022	.022	.023	.023	.024	.024	.024	.025	.026	.026	.026
39	.023	.023	.024	.024	.024	.025	.025	.026	.026	.027	.027	.028	.028	.029	.029
40	.025	.025	.026	.026	.027	.027	.028	.028	.029	.030	.030	.030	.031	.031	.032
41	.027	.027	.028	.029	.029	.030	.030	.031	.031	.032	.033	.033	.034	.034	.035
42	.029	.030	.030	.031	.032	.032	.033	.033	.034	.034	.035	.035	.036	.037	.038
43	.031	.032	.033	.033	.034	.035	.035	.036	.036	.037	.038	.038	.039	.040	.040
44	.033	.034	.035	.035	.036	.037	.038	.038	.039	.040	.040	.041	.042	.042	.043
45	.036	.037	.037	.038	.039	.039	.040	.041	.042	.042	.043	.044	.045	.045	.046
46	.038	.038	.039	.040	.041	.042	.043	.043	.044	.045	.046	.046	.047	.048	.049
47	.040	.041	.042	.042	.043	.044	.045	.046	.047	.048	.048	.049	.050	.051	.052
48	.042	.043	.044	.045	.046	.047	.047	.048	.049	.050	.051	.052	.053	.053	.054
49	.044	.045	.046	.047	.048	.049	.050	.051	.052	.052	.054	.054	.055	.056	.057
50	.046	.047	.048	.049	.050	.051	.052	.053	.054	.055	.056	.057	.058	.059	.060
51	.049	.050	.051	.052	.053	.054	.055	.056	.057	.058	.059	.000	.061	.062	.063
52	.051	.052	.053	.054	.055	.056	.057	.058	.059	.060	.061	.062	.064	.065	.066
53	.053	.054	.055	.056	.057	.058	.060	.061	.062	.063	.064	.065	.066	.067	.068
54	.055	.056	.057	.058	.060	.061	.062	.063	.064	.065	.067	.068	.069	.070	.071
55	.057	.058	.060	.061	.062	.063	.064	.065	.066	.068	.069	.070	.071	.073	.074
56	.060	.061	.062	.063	.064	.065	.067	.068	.069	.070	.072	.073	.074	.075	.077
57	.062	.063	.064	.065	.067	.068	.069	.070	.072	.073	.075	.076	.077	.078	.080
58	.064	.065	.066	.068	.069	.070	.071	.073	.074	.076	.077	.078	.080	.081	.082
59	.066	.068	.069	.070	.072	.073	.074	.075	.077	.078	.080	.081	.083	.084	.085
60	.068	.070	.071	.072	.074	.076	.077	.078	.079	.081	.082	.084	.085	.086	.088
61	.070	.072	.073	.074	.076	.077	.079	.080	.082	.083	.085	.086	.088	.089	.091
62	.073	.074	.076	.077	.079	.080	.082	.083	.085	.086	.088	.089	.091	.092	.094
63	.075	.076	.078	.079	.081	.082	.084	.085	.087	.088	.090	.091	.093	.095	.096
64	.077	.078	.080	.081	.083	.085	.086	.088	.090	.091	.093	.094	.096	.097	.099
65	−.079	−.080	−.082	−.084	−.086	−.087	−.089	−.090	−.092	−.093	−.095	−.097	−.099	−.100	−.102

TABLE III. — REDUCTION OF BAROMETER READING TO 32°. — *Continued.*

Inches.

Temperature.	24.0	24.5	25.0	25.5	26.0	26.5	27.0	27.5	28.0	28.5	29.0	29.5	30.0	30.5	31.0
65	−.079	−.080	−.082	−.084	−.086	−.087	−.089	−.090	−.092	−.093	−.095	−.097	−.099	−.100	−.102
66	.081	.083	.085	.086	.088	.089	.091	.093	.095	.096	.098	.099	.101	.103	.105
67	.083	.085	.087	.088	.090	.092	.094	.095	.097	.099	.101	.102	.104	.106	.108
68	.085	.087	.089	.090	.093	.094	.096	.098	.100	.101	.103	.105	.107	.108	.110
69	.088	.089	.091	.093	.095	.097	.099	.100	.102	.104	.106	.107	.110	.111	.113
70	.090	.092	.094	.096	.097	.099	.101	.103	.105	.106	.109	.110	.112	.114	.116
71	.092	.094	.096	.098	.100	.101	.103	.105	.107	.109	.111	.113	.115	.116	.119
72	.094	.096	.098	.100	.102	.104	.106	.108	.110	.112	.114	.116	.118	.120	.122
73	.096	.098	.100	.102	.104	.106	.108	.110	.112	.114	.116	.118	.120	.122	.124
74	.098	.100	.103	.105	.107	.109	.111	.113	.115	.117	.119	.121	.123	.125	.127
75	.101	.102	.105	.106	.109	.111	.113	.115	.117	.119	.122	.124	.126	.128	.130
76	.103	.104	.107	.109	.111	.113	.116	.118	.120	.122	.124	.126	.128	.130	.133
77	.105	.107	.109	.111	.114	.116	.118	.120	.122	.124	.127	.129	.131	.133	.136
78	.107	.109	.112	.113	.116	.118	.120	.122	.125	.127	.129	.131	.134	.136	.138
79	.109	.111	.114	.116	.118	.120	.123	.125	.127	.129	.132	.134	.137	.139	.141
80	.111	.113	.116	.118	.121	.123	.125	.127	.130	.132	.135	.137	.139	.141	.144
81	.114	.116	.118	.120	.123	.125	.128	.130	.132	.134	.137	.139	.142	.144	.147
82	.116	.118	.121	.122	.125	.128	.130	.132	.135	.137	.140	.142	.145	.147	.149
83	.118	.120	.123	.125	.128	.130	.133	.135	.138	.140	.142	.145	.147	.149	.152
84	.120	.122	.125	.127	.130	.132	.135	.138	.140	.142	.145	.147	.150	.152	.155
85	.122	.124	.127	.129	.132	.134	.137	.139	.143	.145	.148	.150	.153	.155	.158
86	.124	.126	.128	.130	.135	.137	.140	.143	.145	.148	.150	.153	.155	.158	.161
87	.126	.129	.132	.134	.137	.139	.142	.144	.148	.150	.153	.155	.158	.161	.163
88	.129	.131	.134	.137	.139	.142	.145	.147	.150	.152	.155	.158	.161	.163	.166
89	.131	.133	.136	.139	.142	.144	.147	.150	.153	.155	.158	.161	.164	.166	.169
90	.133	.136	.138	.141	.144	.147	.150	.153	.155	.157	.161	.164	.166	.169	.172
91	.135	.138	.141	.143	.146	.149	.152	.155	.158	.160	.163	.166	.169	.172	.175
92	.137	.140	.143	.146	.149	.152	.154	.157	.160	.163	.166	.169	.172	.175	.177
93	.139	.142	.145	.148	.151	.154	.157	.160	.163	.166	.168	.171	.174	.177	.180
94	.142	.145	.147	.150	.153	.156	.159	.162	.165	.168	.171	.174	.177	.180	.183
95	.144	.147	.150	.153	.156	.159	.162	.165	.168	.171	.174	.177	.180	.183	.186
96	.146	.149	.152	.155	.158	.161	.164	.167	.170	.173	.176	.179	.182	.185	.188
97	.148	.151	.154	.157	.160	.164	.167	.170	.173	.176	.179	.182	.185	.188	.191
98	.150	.153	.156	.160	.163	.166	.169	.172	.175	.178	.181	.185	.188	.191	.194
99	.152	.155	.159	.162	.165	.168	.171	.175	.178	.181	.184	.187	.190	.194	.197
100	−.154	−.157	−.161	−.164	−.167	−.171	−.174	−.177	−.180	−.184	−.187	−.190	−.193	−.197	−.200

TABLE IV. — TABLE FOR REDUCING OBSERVATIONS OF THE BAROMETER TO SEA LEVEL, CORRECTION ADDITIVE.

Height, in feet	\-20°	\-10°	0°	10°	20°	30°	40°	50°	60°	70°	80°	90°	100°
10	.013	.013	.012	.012	.012	.012	.011	.011	.011	.011	.010	.010	.010
20	.026	.025	.025	.024	.023	.023	.023	.022	.022	.021	.021	.020	.020
30	.039	.038	.037	.036	.035	.034	.034	.033	.032	.032	.031	.030	.030
40	.052	.050	.049	.048	.047	.046	.045	.044	.043	.042	.041	.040	.040
50	.065	.063	.061	.060	.059	.058	.056	.055	.054	.053	.052	.051	.050
60	.077	.076	.074	.072	.070	.069	.068	.066	.065	.063	.062	.061	.050
70	.090	.088	.086	.084	.082	.081	.078	.077	.076	.074	.072	.071	.060
80	.103	.101	.098	.096	.094	.092	.090	.088	.086	.084	.082	.081	.079
90	.116	.113	.111	.108	.105	.104	.101	.099	.097	.095	.093	.091	.089
100	.129	.126	.123	.120	.117	.115	.112	.110	.108	.105	.103	.101	.099
110	.142	.139	.135	.132	.129	.126	.123	.121	.119	.116	.113	.111	.109
120	.155	.151	.148	.144	.140	.138	.134	.132	.129	.126	.124	.121	.119
130	.168	.164	.160	.156	.152	.149	.146	.143	.140	.137	.134	.131	.129
140	.181	.176	.172	.168	.164	.161	.157	.154	.151	.147	.144	.141	.130
150	.194	.189	.185	.180	.176	.172	.168	.165	.162	.158	.155	.152	.149
160	.206	.201	.197	.192	.187	.183	.179	.176	.172	.168	.165	.162	.158
170	.219	.214	.209	.204	.199	.195	.190	.187	.183	.179	.175	.172	.168
180	.232	.227	.222	.216	.211	.206	.202	.198	.194	.189	.185	.182	.178
190	.245	.239	.234	.228	.222	.218	.213	.209	.204	.200	.196	.192	.188
200	.258	.252	.246	.240	.234	.229	.224	.220	.215	.210	.206	.202	.198
210	.271	.264	.258	.252	.246	.240	.235	.231	.226	.221	.216	.212	.208
220	.284	.277	.270	.264	.257	.252	.246	.242	.236	.231	.227	.222	.218
230	.296	.289	.283	.276	.269	.263	.257	.253	.247	.242	.237	.232	.228
240	.309	.302	.295	.288	.281	.275	.269	.264	.258	.252	.248	.242	.238
250	.322	.314	.307	.300	.293	.286	.280	.275	.269	.263	.258	.253	.248
260	.335	.327	.319	.311	.304	.297	.291	.285	.279	.273	.268	.263	.257
270	.348	.339	.331	.323	.316	.309	.302	.296	.290	.284	.278	.273	.267
280	.360	.352	.344	.335	.328	.320	.314	.307	.301	.294	.288	.283	.277
290	.373	.364	.356	.347	.339	.332	.325	.318	.311	.305	.299	.293	.287
300	.386	.377	.368	.359	.351	.343	.336	.329	.322	.315	.309	.303	.297
310	.399	.389	.380	.371	.363	.354	.347	.340	.333	.326	.319	.313	.307
320	.412	.402	.392	.383	.374	.366	.358	.351	.343	.336	.329	.323	.317
330	.424	.414	.404	.395	.386	.377	.369	.362	.354	.347	.340	.333	.326
340	.437	.427	.416	.407	.397	.389	.380	.373	.365	.357	.350	.343	.336
350	.450	.439	.429	.419	.409	.400	.392	.384	.376	.368	.360	.353	.346
360	.463	.451	.441	.430	.421	.411	.403	.394	.386	.378	.370	.363	.356
370	.476	.464	.453	.442	.432	.423	.414	.405	.397	.389	.380	.373	.366
380	.488	.476	.465	.454	.444	.434	.425	.416	.408	.399	.391	.383	.375
390	.501	.489	.477	.466	.455	.446	.436	.427	.418	.410	.401	.393	.385
400	.514	.501	.489	.478	.467	.457	.447	.438	.429	.420	.411	.403	.395
410	.527	.513	.501	.490	.479	.468	.458	.449	.440	.430	.421	.413	.405
420	.530	.526	.513	.502	.490	.480	.469	.460	.450	.441	.431	.423	.415
430	.552	.538	.525	.513	.502	.491	.480	.470	.461	.451	.442	.433	.425
440	.565	.551	.537	.525	.513	.502	.491	.481	.471	.462	.452	.443	.434
450	.578	.563	.550	.537	.525	.513	.503	.492	.482	.472	.462	.453	.444
460	.590	.575	.562	.549	.537	.525	.514	.503	.493	.482	.472	.463	.454
470	.603	.588	.574	.561	.548	.536	.525	.514	.503	.493	.482	.473	.464
480	.616	.600	.586	.572	.560	.547	.536	.524	.514	.503	.493	.483	.474
490	.628	.613	.598	.584	.571	.559	.547	.535	.524	.514	.503	.493	.483
500	.641	.625	.610	.596	.583	.570	.558	.546	.535	.524	.513	.503	.493
510	.654	.637	.622	.608	.594	.581	.569	.557	.545	.534	.523	.513	.503
520	.666	.650	.634	.620	.606	.593	.580	.568	.556	.545	.533	.523	.513
530	.679	.662	.646	.631	.617	.604	.591	.578	.566	.555	.544	.533	.522
540	.691	.675	.658	.643	.629	.615	.602	.589	.577	.565	.554	.543	.532

METEOROLOGICAL TABLES. 169

TABLE IV. — FOR REDUCING OBSERVATIONS OF THE BAROMETER TO SEA LEVEL. — *Continued.*

Height, in feet.	Temperature of external air — degrees Fahrenheit.												
	−20°	−10°	0°	10°	20°	30°	40°	50°	60°	70°	80°	90°	100°
550	.704	.687	.670	.655	.640	.626	.613	.600	.587	.575	.564	.553	.542
560	.717	.699	.683	.667	.652	.638	.624	.611	.598	.586	.574	.563	.552
570	.729	.712	.695	.679	.663	.649	.635	.622	.608	.596	.584	.573	.562
580	.742	.724	.707	.690	.675	.660	.646	.632	.619	.606	.595	.583	.571
590	.754	.737	.719	.702	.686	.672	.657	.643	.629	.617	.605	.593	.581
600	.767	.749	.731	.714	.698	.683	.668	.654	.640	.627	.615	.603	.591
610	.780	.761	.743	.726	.709	.694	.679	.665	.650	.637	.625	.613	.601
620	.792	.774	.755	.738	.721	.705	.690	.675	.661	.648	.635	.623	.611
630	.805	.786	.767	.749	.732	.717	.701	.686	.671	.658	.645	.633	.620
640	.817	.798	.779	.761	.744	.728	.712	.697	.682	.668	.655	.643	.630
650	.830	.811	.791	.773	.755	.739	.723	.708	.692	.679	.666	.653	.640
660	.843	.823	.803	.785	.767	.750	.734	.718	.703	.689	.676	.662	.650
670	.855	.835	.815	.797	.778	.761	.745	.729	.713	.699	.686	.672	.660
680	.868	.847	.827	.808	.790	.773	.756	.740	.724	.709	.696	.682	.669
690	.880	.860	.839	.820	.801	.784	.767	.750	.734	.720	.706	.692	.679
700	.893	.872	.851	.832	.813	.795	.778	.761	.745	.730	.716	.702	.689
710	.905	.884	.863	.844	.824	.806	.789	.772	.755	.740	.726	.712	.698
720	.918	.896	.875	.855	.836	.817	.800	.782	.766	.751	.736	.722	.708
730	.930	.909	.887	.867	.847	.829	.811	.793	.776	.761	.746	.732	.718
740	.943	.921	.899	.879	.859	.840	.822	.804	.787	.771	.756	.742	.728
750	.955	.933	.911	.891	.870	.851	.833	.815	.797	.782	.767	.752	.738
760	.968	.945	.922	.902	.881	.862	.843	.825	.808	.792	.777	.761	.747
770	.980	.957	.934	.914	.893	.873	.854	.836	.818	.802	.787	.771	.757
780	.993	.970	.946	.926	.904	.885	.865	.847	.829	.812	.797	.781	.767
790	1.005	.982	.958	.937	.916	.896	.876	.857	.839	.823	.807	.791	.776
800	1.018	.994	.970	.949	.927	.907	.887	.868	.850	.833	.817	.801	.786
810	1.030	1.006	.982	.961	.938	.918	.898	.878	.860	.843	.827	.811	.796
820	1.043	1.018	.994	.972	.950	.929	.909	.889	.871	.854	.837	.821	.805
830	1.055	1.031	1.006	.984	.961	.940	.920	.900	.881	.864	.847	.831	.815
840	1.068	1.043	1.018	.995	.973	.951	.931	.911	.892	.874	.857	.841	.825
850	1.080	1.055	1.030	1.007	.984	.962	.942	.922	.902	.885	.867	.851	.835
860	1.093	1.067	1.041	1.019	.995	.974	.952	.932	.913	.895	.877	.860	.844
870	1.105	1.079	1.053	1.030	1.007	.985	.963	.943	.923	.905	.887	.870	.854
880	1.118	1.092	1.065	1.042	1.018	.996	.974	.954	.934	.915	.897	.880	.864
890	1.130	1.104	1.077	1.053	1.030	1.007	.985	.964	.944	.926	.907	.890	.873
900	1.143	1.116	1.089	1.065	1.041	1.018	.996	.975	.955	.936	.917	.900	.883
910	1.155	1.128	1.101	1.077	1.052	1.029	1.007	.986	.965	.946	.927	.910	.893
920	1.168	1.140	1.113	1.088	1.064	1.040	1.018	.996	.976	.956	.937	.920	.902
930	1.180	1.152	1.125	1.100	1.075	1.051	1.029	1.007	.986	.967	.947	.929	.912
940	1.193	1.164	1.137	1.111	1.086	1.062	1.040	1.017	.997	.977	.957	.939	.921
950	1.205	1.177	1.149	1.123	1.098	1.074	1.051	1.028	1.007	.987	.967	.949	.931
960	1.217	1.189	1.160	1.135	1.109	1.085	1.061	1.039	1.017	.997	.977	.959	.941
970	1.230	1.201	1.172	1.146	1.120	1.096	1.072	1.049	1.028	1.007	.987	.969	.950
980	1.242	1.213	1.184	1.158	1.131	1.107	1.083	1.060	1.038	1.018	.997	.978	.960
990	1.255	1.225	1.196	1.169	1.143	1.118	1.094	1.070	1.049	1.028	1.007	.988	.969
1000	1.267	1.237	1.208	1.181	1.154	1.129	1.105	1.081	1.059	1.038	1.017	.998	.979
1010	1.279	1.249	1.220	1.192	1.165	1.140	1.116	1.092	1.069	1.048	1.027	1.008	.989
1020	1.292	1.261	1.232	1.204	1.177	1.151	1.127	1.102	1.080	1.058	1.037	1.018	.998
1030	1.304	1.273	1.243	1.215	1.188	1.162	1.137	1.113	1.090	1.069	1.047	1.027	1.008
1040	1.317	1.285	1.255	1.227	1.199	1.173	1.148	1.123	1.101	1.079	1.057	1.037	1.017
1050	1.329	1.298	1.267	1.238	1.211	1.184	1.159	1.134	1.111	1.089	1.067	1.047	1.027
1060	1.341	1.310	1.279	1.250	1.222	1.195	1.170	1.145	1.121	1.099	1.077	1.057	1.037
1070	1.354	1.322	1.291	1.261	1.233	1.206	1.181	1.155	1.132	1.109	1.087	1.066	1.046
1080	1.366	1.334	1.302	1.273	1.244	1.217	1.191	1.166	1.142	1.120	1.097	1.076	1.056
1090	1.379	1.346	1.314	1.284	1.256	1.228	1.202	1.176	1.153	1.130	1.107	1.086	1.065

TABLE IV. — FOR REDUCING OBSERVATIONS OF THE BAROMETER TO SEA LEVEL. — *Continued.*

Height, in feet.	Temperature of external air — degrees Fahrenheit.												
	—20°	—10°	0°	10°	20°	30°	40°	50°	60°	70°	80°	90°	100°
1100	1.391	1.358	1.326	1.296	1.267	1.239	1.213	1.187	1.163	1.140	1.117	1.096	1.075
1110	1.403	1.370	1.338	1.307	1.278	1.250	1.224	1.198	1.173	1.150	1.127	1.106	1.085
1120	1.416	1.382	1.350	1.319	1.289	1.261	1.235	1.208	1.184	1.160	1.137	1.115	1.094
1130	1.428	1.394	1.361	1.330	1.301	1.272	1.245	1.219	1.194	1.170	1.147	1.125	1.104
1140	1.440	1.406	1.373	1.342	1.312	1.283	1.256	1.229	1.204	1.180	1.157	1.135	1.113
1150	1.453	1.418	1.385	1.353	1.323	1.294	1.267	1.240	1.215	1.191	1.167	1.145	1.123
1160	1.465	1.430	1.397	1.365	1.334	1.305	1.278	1.251	1.225	1.201	1.177	1.154	1.133
1170	1.477	1.442	1.409	1.376	1.345	1.315	1.289	1.261	1.235	1.211	1.187	1.164	1.142
1180	1.489	1.454	1.420	1.388	1.357	1.327	1.299	1.272	1.245	1.221	1.197	1.174	1.152
1190	1.502	1.466	1.432	1.399	1.368	1.338	1.310	1.282	1.256	1.231	1.207	1.183	1.161
1200	1.514	1.478	1.444	1.411	1.379	1.349	1.321	1.293	1.266	1.241	1.217	1.193	1.171
1210	1.526	1.490	1.456	1.422	1.390	1.360	1.332	1.303	1.276	1.251	1.227	1.203	1.180
1220	1.539	1.502	1.467	1.434	1.401	1.371	1.342	1.314	1.288	1.261	1.237	1.212	1.190
1230	1.551	1.514	1.479	1.445	1.413	1.382	1.353	1.324	1.297	1.271	1.247	1.222	1.199
1240	1.563	1.526	1.491	1.457	1.424	1.393	1.364	1.335	1.307	1.281	1.257	1.232	1.209
1250	1.576	1.538	1.502	1.468	1.435	1.404	1.374	1.345	1.317	1.291	1.266	1.242	1.218
1260	1.588	1.550	1.514	1.479	1.446	1.415	1.385	1.356	1.328	1.302	1.276	1.251	1.228
1270	1.600	1.562	1.526	1.491	1.457	1.426	1.396	1.366	1.338	1.312	1.286	1.261	1.237
1280	1.612	1.574	1.538	1.502	1.469	1.437	1.407	1.377	1.348	1.322	1.296	1.271	1.247
1290	1.625	1.586	1.549	1.514	1.480	1.448	1.417	1.387	1.359	1.332	1.306	1.280	1.256
1300	1.637	1.598	1.561	1.525	1.491	1.459	1.428	1.398	1.369	1.342	1.316	1.290	1.266
1310	1.649	1.610	1.573	1.536	1.502	1.470	1.439	1.408	1.379	1.352	1.326	1.300	1.275
1320	1.661	1.622	1.584	1.548	1.513	1.481	1.449	1.419	1.390	1.362	1.336	1.309	1.285
1330	1.674	1.634	1.596	1.559	1.525	1.492	1.460	1.429	1.400	1.372	1.346	1.319	1.294
1340	1.686	1.646	1.608	1.571	1.536	1.503	1.471	1.440	1.410	1.382	1.356	1.329	1.304
1350	1.698	1.658	1.620	1.582	1.547	1.514	1.482	1.450	1.420	1.393	1.366	1.339	1.313
1360	1.710	1.669	1.631	1.593	1.558	1.524	1.492	1.461	1.431	1.403	1.375	1.348	1.323
1370	1.722	1.681	1.643	1.605	1.569	1.535	1.503	1.471	1.441	1.413	1.385	1.358	1.332
1380	1.735	1.693	1.655	1.616	1.581	1.546	1.514	1.482	1.451	1.423	1.395	1.368	1.342
1390	1.747	1.705	1.666	1.628	1.592	1.557	1.524	1.492	1.462	1.433	1.405	1.377	1.351
1400	1.759	1.717	1.678	1.639	1.603	1.568	1.535	1.503	1.472	1.443	1.415	1.387	1.361
1410	1.771	1.729	1.690	1.650	1.614	1.579	1.546	1.513	1.482	1.453	1.425	1.397	1.370
1420	1.783	1.741	1.701	1.662	1.625	1.590	1.556	1.524	1.492	1.463	1.435	1.406	1.380
1430	1.796	1.753	1.713	1.673	1.636	1.601	1.567	1.534	1.503	1.473	1.444	1.416	1.389
1440	1.808	1.765	1.724	1.685	1.647	1.612	1.577	1.545	1.513	1.483	1.454	1.426	1.399
1450	1.820	1.777	1.736	1.696	1.658	1.623	1.588	1.555	1.523	1.493	1.464	1.436	1.408
1460	1.832	1.788	1.748	1.707	1.670	1.633	1.599	1.565	1.533	1.503	1.474	1.445	1.418
1470	1.844	1.800	1.759	1.719	1.681	1.644	1.609	1.576	1.543	1.513	1.484	1.455	1.427
1480	1.857	1.812	1.771	1.730	1.692	1.655	1.620	1.586	1.554	1.523	1.493	1.465	1.437
1490	1.869	1.824	1.782	1.742	1.703	1.666	1.630	1.597	1.564	1.533	1.503	1.474	1.446
1500	1.881	1.836	1.794	1.753	1.714	1.677	1.641	1.607	1.574	1.543	1.513	1.484	1.456

APPENDIX A.

SUGGESTIONS TO TEACHERS.

It is the object of this book to lead the student to the independent discovery of the most important facts in our ordinary weather conditions, and of the interrelations of the different weather elements. This practical study having taught something as to the real nature of atmospheric phenomena by actual observation, rapid and substantial progress may be made in the knowledge of the distribution and of the explanation of similar phenomena in other parts of the world, as derived through a study of the text-books. By means of this combination of the two kinds of study, the inductive and the didactic, the advantages of both may be preserved, and the slow progress of the first method and the unsound progress of the second may be avoided. This book is not a text-book, and it therefore does not attempt to give explanations of various phenomena discovered by the class. Explanations will, of course, be called for by the scholars, in increasing number as the work progresses, and the larger relations of the study become apparent. It is best, if possible, to leave the more complicated matters (such as the cause of the deflection of the wind from the gradient, of cyclones and anticyclones, etc.) until the subjects can be taken up in detail and fully explained, for instance in the later years of the high school course. It is not advisable to raise such complicated questions in the grammar school work if they can be avoided. The teacher who has a fairly good knowledge of one comprehensive modern text-book of meteorology, such as Davis's *Elementary Meteorology*, will find himself sufficiently well equipped to answer the questions put by the class.

The value of the work outlined in this little book can be much increased if the larger applications of the lessons here learned are strongly emphasized. Suggestions along this line have been made in fine print throughout the text, but the examples given may be further extended to the great advantage of the student. Careful attention ought to be given to the formulating and writing out of the generalizations reached by the class, for in these written summaries the results are preserved in compact form.

CHAPTER I.

The work outlined in this chapter is adapted to the lower grades in the grammar school. It is assumed that the pupils have already had some preliminary training in the simplest non-instrumental weather observations, such as can readily be made during the primary school years. For the convenience of teachers who may desire it, a brief outline of work suited to the primary school grades is here given. It is desirable that even older scholars be given some such training as this before they take up the exercises of Chapter II.

The central idea in this elementary work is to train the children in intelligent weather observation, so that they may come to appreciate what our typical weather changes are; that they may recognize the types as they recur, and may see how each example differs from, or accords with, those that have preceded it. We are all so directly affected by the weather conditions prevailing at any time that even the youngest children are forced, unconsciously to be sure, to take some notice of these changes. The work of the teacher is, therefore, simply to direct attention to what is already seen.

When the children come to school on some snowy winter day, with a northeast wind, chilling and damp, attention may be called to the need of overshoes and overcoats, to the piling up of the snow in deep drifts at certain places near the school or in the town, while in other places the ground is left bare; to the ease with which snowballs may be made, and to other facts which will very readily suggest themselves. A day or two after such a storm, when the sun is shining bright in a cloudless sky, when there is no wind and the air is dry, cold, and crisp, the contrasts between these two weather types should be brought out. Instead of snow we now have sunshine; instead of a damp, chilling northeaster we now have a calm and the air is dry; snowballs cannot easily be made in the early morning because the snow is frozen hard and is too dry, but towards noon, if the temperature rise high enough, there may be thawing on the tops or sides of the snowdrifts, and there the snow becomes soft enough for snowballing. Another weather type, often noted during our winter in the central and eastern United States, and strongly contrasted with both of the preceding conditions, is that which brings us a warm, damp, southerly wind, frequently accompanied by heavy rains. As these damp winds blow over snow-covered surfaces they become foggy and the ground is said to "smoke"; the heavy rain rapidly melts the snow; slush and mud make bad walking; rivers and brooks rise rapidly, perhaps overflowing their banks; low-lying places become filled with standing water. These and other features should all be brought out by the teacher, not by telling the class of them directly, but by judicious questioning, and they should be contrasted with the conditions which may immediately follow, when the storm has cleared off, and when the low temperatures brought by a cold wave, with its dry northwest wind, have resulted in freezing lakes, rivers, and brooks, and when skating and sliding may be indulged in. Early summer weather conditions, with their characteristic warm spells, cumulus clouds, thunderstorms, and (near the coast) sea breezes, furnish another long list of typical changes that should be just as carefully noted and described as the more striking winter characteristics. Autumn types add further to the list, which might be extended almost indefinitely.

One whole year of the grammar school course may well be given to the observations suggested in Chapter I, provided that there is no need of hastening on to the more advanced work. The advantage of extending the course over a whole school year is great, because such extension gives opportunity for becoming familiar with late summer, autumn, winter, spring, and early summer weather types, and this is far better than attempting to crowd all the work into one short season. The interest of a class can easily be kept up throughout a school year by means of a progressive system of observations. It is best to vary the observations from time to time, and to arrange them so that, beginning with the more simple, they shall gradually become more complete and more advanced as the year goes on. Thus, starting with temperature observations alone, these may be continued for one or two weeks before they are supplemented by records of wind direction and velocity. After some practice in the observation of these two weather elements (say during one month), data as to the state of the sky may be added. Cloud observations themselves may well be graded during successive weeks, so that, beginning with the simplest notes concerning amounts of cloudiness, the pupils shall gradually advance to the point of observing, and perhaps even of sketching, the common cloud forms and their changes. Thus an important step will have been taken towards appreciating the need of a standard cloud classification, which may be given later.

The addition of records of precipitation completes the list of simple non-instrumental weather observations, and these records, as well as the cloud records, can easily be graded, so that, during successive weeks, every week's work shall be different from that of every other week. In this progression from the simpler to the more complicated observations lies the secret of making the work attractive. Nothing will sooner check interest in the study than the necessity of making exactly the same observations day after day and week after week throughout the year. A graded course of non-instrumental observation, such as is suggested, gives a very practical general knowledge of our common weather types and changes, and of the relations of one weather element to another. The questions asked under the different headings in this chapter are designed to awaken the interest of the scholars, and to call their attention to the more important points of diurnal, cyclonic, and seasonal changes in weather elements. The teacher will readily think of other questions which may be suggested for the consideration of the class.

Although the non-instrumental records are of little value for future reference, as compared with the instrumental observations, they should nevertheless be systematically preserved by the class in their record books. After discussion of the daily observations made by the different scholars, or by one of their number, the records may be written upon one of the blackboards reserved for this purpose. At the close of the day, or the next morning, the blackboard notes should be entered in a record book kept in the schoolroom. The teacher may guide in the discussion of the observations; may suggest points overlooked by the scholars; may draw comparisons between the weather conditions of other weeks and of other days. This talking over of the observations is most important, as it never fails to bring out much of interest.

CHAPTER II.

This work may usually be begun in the early years of the grammar school course, as soon as the non-instrumental observations have been satisfactorily completed. The scheme of progressive observations already suggested may be followed to advantage in the instrumental work as well as in the non-instrumental. It is often a good plan to have a different scholar assigned to the task of taking the observations every day, or it may be more advisable to divide the work, making one responsible for the temperature observations, another for the precipitation, etc. It is well to have the daily instrumental weather records written upon the blackboard in the schoolroom, as already suggested in the case of the non-instrumental observations. At the end of each day the blackboard data should be entered in a permanent record book by some one of the scholars, and some ingenuity can be exercised in devising the best scheme for keeping this record. The record book should be carefully preserved in the schoolroom, where it may be referred to by the scholars of future years when any unusually severe storm, or a spell of excessively hot or dry weather, or a remarkable cold wave occurs, in order that comparison with past occurrences of a similar kind may be made. It is well to have the record book of large size, and to have each day's record entered across two full pages. On the left-hand page the temperature, pressure, rainfall, wind direction and velocity, etc., may be entered, each observation in its proper column, the number of columns being increased according to the increasing number of observations. The right-hand page may be left for "Remarks." These "Remarks" should include notes of any meteorological phenomena which did not find a place in the columns reserved for the regular observations, *e.g.*, occurrence of hail, or frozen rain ; damage by lightning, winds, or floods; freezing up of rivers or brooks; interruption of railroad or street-car traffic by snow, etc., and, in general, explanatory comments on the weather conditions. Instructive lessons may be taught as to the relation of the local weather conditions which prevail in the vicinity of the school, and those of other portions of the country, by comments on newspaper despatches concerning gales along the coast or on the lakes, and resulting damage to shipping; of snow blockades and stalled trains ; of severe thunderstorms and tornadoes; of hot waves and sunstrokes, or of cold waves and the destruction of crops or fruits by the frost. The scholars should be encouraged to bring into the class any comments on such phenomena as may be of interest in the work. Such of these newspaper clippings as are of the most value may be pasted in the space reserved for the " Remarks," where they may be referred to by succeeding classes; and in this space also may be pasted at the end of each week the barograph and thermograph sheets, if these instruments are in use at the school.

CHAPTER III.

These observations may usually be profitably undertaken in the later grammar and in the high school years. The instruments described, while all desirable, are by no means all necessary, and no teacher should postpone the establishment of a course in observational meteorology for the reason that a complete set of first-class instruments cannot be secured at the start.

If the school is provided with a psychrometer, there will be no need of the ordinary thermometer, because the psychrometer gives the true air temperature. It is well, however, to have both stationary wet and dry bulb thermometers, in the shelter, for ordinary school use, and also a sling psychrometer for use in the meteorological *field work* which forms an important part of the more advanced instrumental work in meteorology. The sling psychrometer may, of course, be used simply as an ordinary sling thermometer.

The simple form of mercurial barometer, without vernier and without attached thermometer, described in Chapter II, will be found the best barometer for general school use. The standard barometer, described in this chapter, is too expensive and too complicated to come into extended use in our schools. Full instructions concerning the care, the reading, and the corrections of the standard mercurial barometer are published by the Weather Bureau, and to these instructions teachers who have such an instrument are referred. (See Appendix B.)

The form of table given at the end of this chapter is intended merely as a suggestion, and not as a rigid scheme to be adopted in every school. In using the instruments here described, practice with the maximum and minimum thermometers (in addition to the simpler work of Chapter II) may be given before any attempt is made to have the class use the psychrometer. And in using the psychrometer one week may well be given to the determination of the dew-point alone, before the wet and dry bulb readings are employed to determine the relative humidity. Absolute humidity, which is not referred to in this chapter, may, if the teacher deem it advisable, be added as another weather element for study. A refinement in the notes on the state of the sky is suggested, viz., that cloudiness should be recorded in tenths of the sky covered by clouds. This is an advance over the earlier, less accurate cloud observations, and is in line with such a progressive scheme as has been recommended. This book is not intended to present a rigid scheme of observational work in meteorology, alike for all schools, but rather to make suggestions for the guidance of teachers in laying out such a course as may fit their own cases.

Under the heading *Summary of Observations* only a few of the most important climatic elements have been noted. The list may easily be extended by the addition of such data as the following: For temperature, mean diurnal range; mean diurnal variability (the mean of the differences between the successive daily means). For humidity, monthly mean absolute humidity. For precipitation, the maximum daily precipitation; the number of rainy and snowy days in every month, the number of clear, fair, and cloudy days in every month; the mean frequency of rainfall in every month (number of rainy days divided by the total number of days); the number of days with thunderstorms, etc.

It is important that the monthly summaries should be discussed in the class, and that the scholars should give verbal statements as to the numerical results which they have obtained. In this way the work will have a living interest, which the mere compilation of summaries does not possess.

CHAPTER IV.

The first thing for any teacher to do who intends to establish a course in meteorology is to secure a supply of daily weather maps. Arrangements should be made to have them mailed regularly from the nearest map-publishing station of the Bureau. It is important that the Saturday morning map, which is usually not sent to schools, should be included in the set, as the break of two days (Saturday and Sunday) in every week seriously interferes with the value of the work that may be done on consecutive maps. The maps should be securely fastened up in the schoolroom or in the hall. It is advisable to keep at least two maps thus on view all the time, in order that the scholars may be able to study the changes from day to day by comparing the last two or three maps with one another. As soon as they are removed from the wall, the maps should be carefully filed away for future reference. They may be conveniently kept in stiff brown paper folders, each month's maps being enclosed in a separate folder, with the name of the month and the year written on the outside. It is a good plan to keep with the file of maps any newspaper clippings referring to notable meteorological phenomena associated with the conditions shown on the maps. Thus, newspaper accounts of the damage done by a hurricane along the Atlantic Coast; of the blockades caused by a heavy snowstorm in the Northwest; of tornadoes in the Mississippi Valley; of hot waves and sunstrokes in our larger cities, will serve to enliven the study of the maps, and will also help, in later references to them, to recall interesting points that might otherwise escape the memory. It is true that newspapers are prone to exaggerate, and that they are lamentably inaccurate in their use of meteorological terms: but nevertheless they may often be profitably used in such general studies as these. Besides the complete weather map, the school will need a supply of blank weather maps, as used by the Weather Bureau. These may usually be secured from the nearest map-publishing station at cost price. In the case of the preparation of illustrations for permanent class use, as suggested later, it is advisable to employ the blank maps used as the base of the Washington daily weather maps, and to be obtained, at cost price, on application to the Chief of the Weather Bureau, Washington, D. C. These are larger in size ($16\frac{1}{2}$ by $23\frac{3}{4}$ inches) than the station maps, and the paper is of a better quality. Colored illustrations with these Washington blank maps as the base furnish an economical, simple, and effective means of teaching elementary meteorology.

CHAPTER V.

In this chapter a series of six consecutive weather maps is taken as the basis of the work. The study of the weather elements on such a series of maps gives a far clearer understanding of the distribution of these elements and of their relation one with another, than if a far larger number of single maps are studied which do not follow one another in regular sequence. Teachers should add to the map-drawing exercises by giving their classes the data from other sets of maps selected from the school files. Summer maps as well as winter ones should be used, in order that too much emphasis may not be laid on winter conditions. The search for the various cities in which Weather Bureau stations are established, involved in the work of entering the data on the blank maps, furnishes excellent practice in geography. This exercise may be varied, and practice in the location of the different States may be given, if the teacher reads out to the class the temperatures in different parts of the various States, as, *e.g.*, central Arizona, 34°; southwestern Tennessee, 30°; northern Nevada, 38°, etc. As the State names are not given on the Weather Bureau maps, such a method as this will give a very desirable familiarity with the relative positions of the States of the Union. If the school possesses a blackboard outline map of the United States, it may be a good plan, if the class is small, to have one of the pupils enter the temperatures and draw the isotherms on the board before the class, and to let the others correct him if they see that he is going wrong. As to irregularities in the isotherms, which may cause trouble, the officials of the Weather Bureau vary somewhat among themselves in dealing with such cases, and no definite rules can be laid down to fit all occasions. It is best to select maps with few irregularities at first, and in time experience will show how the exceptional cases may be treated.

By the scheme of coloring isothermal maps, suggested in this chapter, a valuable series of permanent illustrations of noteworthy weather types for class use can readily be prepared at very slight cost. For the purposes of these colored illustrations it is best to use the large scale Washington blank weather maps, as suggested in the preceding chapter, and to have each map mounted on heavy cardboard after it has been colored. By means of this mounting the maps are prevented from tearing, and can be kept smooth and in good condition.

The scheme of coloring may be varied to suit the fancy of the scholars, for the preparation of these permanent illustrations may well be intrusted to those of the class who are especially interested in the work, and who are skillful in their use of the paint brush. The work is really very simple and needs only ordinary care. As soon as the drawing of isotherms and the coloring of isothermal charts have been sufficiently studied, the teacher should hang up the daily weather map each day (or a blank map with the isotherms for the day drawn on it), and should call attention to the temperature changes from day to day. In this way the facts of actual temperature changes experienced by the class will be associated with the larger temperature changes shown on the weather maps.

The sections on temperature gradients may be postponed, if the teacher deems it advisable, until other matters of a simpler nature are passed. The idea of rates of change is not an easy one for students to grasp, and it is far better to postpone the consideration of this subject than to involve the class in any confusion at this stage. It is a mistake, however, to omit these sections altogether, as a clear conception of the principle of rates is a valuable part of a student's mental equipment. It is a good plan, in the exercises on lines of temperature decrease, to have maps prepared with *faint* isotherms and *heavy* lines of temperature decrease, in order to emphasize the idea of change of temperature, in contrast to the idea of constancy of temperature expressed by the isotherms.

CHAPTER VI.

In the work on the wind charts it is essential to proceed very slowly, in order that the best results may be obtained by the pupils. Some of the aberrant wind courses, which complicate the discovery of the cyclonic and anticyclonic spirals, may well be omitted in the case of the younger classes, and considerable assistance in facilitating these discoveries may be given by suggestions as to adding intermediate dotted wind arrows, in sympathy with the observed wind directions. The anticyclonic systems are always much more difficult to discover than the cyclonic, and care should be taken to assist the class in this matter as much as seems necessary.

The questions in the text are merely suggestive, and are by no means as numerous as it would be well to make them in the class. The discovery of the spirals will probably be made by degrees. The concise formulation of the facts discovered will furnish excellent basis for exercises in writing.

It is interesting to note that the discussion as to whether the winds blow circularly around, or radially in towards, centers of low pressure, which usually comes up in every class in meteorology at some stage in the study, was carried on in a very animated way about the middle of the present century in this country. Two noted American meteorologists, Redfield and Espy, and their respective followers, took opposite sides in this controversy, Redfield maintaining at the start that the winds moved in circles, and Espy maintaining that they followed radial courses. The truth lay between the two.

CHAPTER VII.

The study of atmospheric pressure is not easy, because the pupils cannot perceive the changes in pressure from day to day by their unaided senses. Especially difficult does this study become if the class has not already had some practice in making barometer readings. When this observational work has not preceded the consideration of the isobaric lines of the daily weather maps, the teacher should introduce the subject of pressure very carefully. The experiment with the Torricellian tube will show the class something of the reality of atmospheric pressure, and the variations in this pressure from day to day can readily be made apparent by means of a few barometer readings, if time cannot be spared for a regular and continued series of barometer observations. A word may be said as to the correction of barometric readings for local influences, in order to make these readings comparable, if this matter has not been previously met with, but care should be taken not to confuse the younger pupils too much with explanations at this stage of the work. It may be well to omit this point unless it is brought up by some pupil. The questions asked in the text are merely suggestive. They may be added to and varied at the discretion of the teacher.

Lines of pressure-decrease should be drawn on all isobaric charts studied in the class, as they are highly instructive. When the isobars are near together, these lines of pressure-decrease may be drawn heavier, to indicate a steeper gradient. The convergence of these lines towards regions of low pressure, and their divergence from regions of high pressure, seen on every map on which these gradients are drawn, emphasizes an important lesson. Before measuring rates of pressure-decrease by means of a scale, considerable practice should be given in the study, by means of the eye, of the rapidity or slowness of decrease of pressure, as shown by the heavier or lighter lines of pressure-decrease. When the broad facts of differing rates are comprehended, then the actual measurement of these rates is a comparatively easy matter. In any case, however, an appreciation of these rates of change is not always readily gained, even by older scholars in the high school. It is, therefore, of prime importance to proceed very slowly indeed at this point, and to have every step fully understood before another step is taken.

The instructions in the text for measuring rates of pressure-decrease are, that these rates shall be recorded as so many hundredths of an inch of change of pressure in one latitude-degree. This is done for the sake of simplicity. If this rate is expressed in hundredths of an inch of pressure in a quarter of a latitude-degree of distance, the numerical value is the same as if expressed in millimeters of pressure per latitude-degree of distance.

CHAPTER VIII.

The fact of the prevalence of different kinds of weather over the country at the same time is of great importance. It should be strongly emphasized by the teacher in the course of the discussion of the maps of weather distribution. Additional exercises of the same sort may be given to advantage, by letting the class plot and study the weather signs taken from any current weather map. Instructive lessons may be taught by talking over, in the class, the different ways in which people all over the country are affected by the character of the weather that happens to prevail where they are.

CHAPTERS IX–XVIII.

The correlation exercises will, as a whole, teach few entirely new facts to the brighter scholars who have faithfully completed the preceding work in observations and in the construction and study of the daily weather maps. These exercises do, however, lead to detailed examination and to the careful working out of the relations which may have been previously noticed in a general way only. They give the repeated illustration which is necessary in order to impress firmly on the mind the lesson that the weather map has to teach.

It is a good plan to let different scholars work out the problems for different months. The results reached in each case should be discussed in the class, and thus each member may have the double advantage of working out his own problem, and of profiting by the work done by his fellows. Throughout these exercises care should be taken to have weather maps of all months studied. The exercise on the correlation of the velocity of the wind with the pressure cannot be undertaken unless the work on temperature and pressure gradients (Chapters V and VII) has been completed.

CHAPTERS XX-XXV.

It is not expected that any one scholar can accomplish all that is here outlined. Examples may be selected from the list, as opportunity offers, so that each scholar shall become familiar with several problems.

Few of the problems suggested call for continuous routine observation at fixed hours. They require, on the other hand, an intelligent examination of ordinary weather phenomena, with special reference to discovering their explanation. In most of the problems a small number of observations will suffice. Under the supervision of the teacher, different problems may be assigned to the several members of a class; or several scholars may work on different parts of the same problem, exchanging records in order to save time. All the scholars should have a general knowledge of the results which have been obtained from the observations made by the other members of their class. The teacher will use his discretion in arranging the order of the problems, and in selecting those that are best suited to the season in which the work is done, to the locality in which the school is situated, and to the facilities and apparatus at command. Although the variety of accessible problems is less in city schools than in country schools, much may be done in the city as well as in the country.

The opportunities for carrying out such observational work vary so much in different schools that it is impossible to give specific instructions, which shall be available in all cases. Some general suggestions are therefore given, which the teacher may supplement by more detailed instructions framed to fit the particular circumstances of each case.

A review of the headings of the different problems shows that a very general correlation exists among them, whereby the subjects of every heading are associated with those of nearly every other.

In other words, every weather element is treated as a function of several other elements. It follows from this that the variety of work here outlined is more apparent than real, and that many problems which appear from their wording to be entirely new are in large part rearrangements of problems previously encountered.

APPENDIX B.

THE EQUIPMENT OF A METEOROLOGICAL LABORATORY.

A. INSTRUMENTS.

Exposed Thermometer (United States Weather Bureau pattern), with brass support, $2.75.

Maximum and Minimum Thermometers (United States Weather Bureau pattern), mounted together on one board, $6.25.

Wet and Dry Bulb Thermometers (United States Weather Bureau pattern), mounted on one board, complete with water cup, $6.50.

Sling Psychrometer (designed by Professor C. F. Marvin, of the United States Weather Bureau), consisting of two exposed mercurial thermometers, mounted on an aluminum back, and provided with polished, turned hard-wood handle and brass trimmings, $5.00.

Sling Psychrometer, consisting of two cylindrical bulb thermometers, mounted one a little above the other upon a light brass frame, with a perforated guard to protect the bulbs while swinging, but which can be raised (by sliding upon the frame) for the purpose of moistening the linen covering of the wet bulb. Much less liable to be broken than the Weather Bureau pattern. $5.00.

Rain Gauge (United States Weather Bureau standard), 8 inches in diameter, complete, with measuring stick, $5.25.

Rain Gauge, 3 inches in diameter, with overflow and measuring stick, $1.25.

Wind Vane (United States Weather Bureau pattern), $10.00.

Anemometer (United States Weather Bureau pattern), with indicator, aluminum cups, and electrical attachment, $25.00.

The same, with painted brass cups, $23.00.

APPENDIX B. 187

Anemometer Register (United States Weather Bureau pattern), with pen and ink attachment, $35.00.

The same, with pencil attachment (old style), $24.00.

Aneroid Barometer (for meteorological work), $14.00-$16.00.

NOTE. — Much cheaper aneroids can be purchased, and may be used to some advantage in the simpler observations in schools.

Mercurial Barometer (Standard United States Weather Bureau pattern), complete with attached thermometer, vernier, etc., $30.00 - $33.00.

NOTE. — The above instruments, as used by the United States Weather Bureau, are made by H. J. Green, 1191 Bedford Avenue, Brooklyn, N. Y. The prices are those given in Green's latest catalogue.

Mercurial Barometer. New improved form, especially designed for school use. Mounted on mahogany back. Scale engraved on aluminum. Divisions of scale on metric and English systems. No vernier, $5.75.

(L. E. Knott Apparatus Co., 14 Ashburton Place, Boston, Mass.)

Thermograph (designed by Dr. Daniel Draper, of New York). Consists of a bimetallic thermometer in a case which carries a disk, with a chart upon its axle instead of hands like the ordinary clock. A pen (resting on the face of the disk) registers the fluctuations of temperature as the chart is carried around. Sizes, 14 x 20 inches, $30.00; 10 x 14 inches, $15.00. This instrument may be purchased of H. J. Green.

Thermograph. Self-recording thermometer (as adopted by the United States Weather Bureau), made by Richard Frères, of Paris. Records continuously on a sheet of paper wound around a revolving drum, which is driven by clock-work inside. Standard size (without duty), $30.00.

Barograph. Self-recording barometer (as adopted by the United States Weather Bureau), made by Richard Frères, of Paris. Similar in general arrangement to the thermograph. Standard size (without duty), $27.60.

These last two instruments can be procured through Glaenzer Frères & Rheinboldt, 26 & 28 Washington Place, New York City.

Instrument Shelter (standard United States Weather Bureau pattern) will hold a set of maximum and minimum thermometers, psychrometer, and a thermograph. May be set up on top of posts driven into the ground, or may be attached to a wall, $18.00.

Barometer Box, for the standard mercurial barometer. Made of mahogany, with glass panels on front and sides; lock and key, and with fittings complete, $8.50.

These may be purchased of H. J. Green.

B. TEXT-BOOKS.

The Story of the Earth's Atmosphere. DOUGLAS ARCHIBALD. New York, D. Appleton & Co., 1898. 18mo, pp. 194. 40 cents.

To be recommended to the general reader who wishes to gain some knowledge of meteorology quickly. Not a text-book. Contains a chapter on " Flight in the Atmosphere."

Elementary Meteorology. WILLIAM MORRIS DAVIS. Boston, Ginn & Co., 1898. 8vo, pp. 355. $2.50.

The most complete of the modern text-books, and the best adapted for use in the systematic teaching of meteorology. The modern views are presented clearly and without the use of mathematics. Portions of it are somewhat too advanced for school study, but teachers will find it invaluable as a reference book in directing the laboratory work, and in answering the questions of school classes.

A Popular Treatise on the Winds. WILLIAM FERREL. New York, John Wiley & Sons, 1890. 8vo, pp. 505. $3.40.

This can hardly be regarded as a *popular* treatise. It embodies, in condensed and chiefly non-mathematical form, the results of Ferrel's researches during his long and profound study of the general circulation and phenomena of the atmosphere. Teachers who advance far into meteorology will find this book indispensable. It is not at all suited for general class-room use.

American Weather. A. W. GREELY. New York, Dodd, Mead & Co., 1888. 8vo, pp. 286. Out of print, but secondhand copies are probably obtainable.

Deals, as the title implies, especially with the weather phenomena of the United States. Contains brief accounts of individual hot and cold waves, hurricanes, blizzards and tornadoes, and gives specific data concerning maxima and minima of temperature, precipitation, etc., in the United States.

Meteorology: Practical and Applied. JOHN WILLIAM MOORE. London, F. J. Rebman, 1894. 8vo, pp. 445. 8 shillings.

A readable book. Considerable space is given to instrumental meteorology. Contains chapters on the climate of the British Isles and on the relations of weather and disease in the British Isles. Especially adapted for the use of English readers.

Elementary Meteorology. ROBERT H. SCOTT. International Scientific Series. London, Kegan Paul, Trench & Co., 1885; Boston, A. A. Waterman & Co., 1889. 8vo, pp. 410. 6 shillings.

The standard text-book in Great Britain. The author is secretary to the Meteorological Council of the Royal Society. Fairly complete, but now somewhat out of date in some portions. It is a useful book in a meteorological library, but does not treat the subject in a way very helpful to the teacher.

Meteorology. THOMAS RUSSELL. New York, The Macmillan Company, 1895. 8vo, pp. 277.

Brief and incomplete as a text-book of meteorology, but containing a very comprehensive account, fully illustrated, of rivers and floods in the United States, and their prediction.

Elementary Meteorology. FRANK WALDO. New York, American Book Company, 1896. 8vo, pp. 373. 90 cents.

A compact summary. Useful to teachers as a handy reference book.

Modern Meteorology. FRANK WALDO. New York, Charles Scribner's Sons, 1893. 8vo, pp. 460. $1.25.

Very complete account of meteorological apparatus and methods, and admirable summary of recent German studies of the thermodynamics and general motions of the atmosphere.

C. INSTRUCTIONS IN THE USE OF INSTRUMENTS.

Instructions for Voluntary Observers. 1899. 8vo, pp. 23. Brief instructions for taking and recording observations of temperature and precipitation with ordinary and maximum and minimum thermometers and with the rain gauge.

APPENDIX B.

Barometers and the Measurement of Atmospheric Pressure. C. F. MARVIN. 1894. 8vo, pp. 74. A pamphlet of information respecting the theory and construction of barometers in general, with summary of instructions for the care and use of the standard Weather Bureau instruments.

Instructions for Obtaining and Tabulating Records from Recording Instruments. 1898. 8vo, pp. 31. Contains directions concerning the care and use of the Richard thermograph and barograph.

NOTE.—These pamphlets are prepared under the direction of Professor Willis L. Moore, Chief of the United States Weather Bureau, and are published, under authority of the Secretary of Agriculture, by the Weather Bureau. They will be found the best guides in making observations, the care of instruments, etc.

D. JOURNALS, ETC.

Monthly Weather Review. Prepared under the direction of Willis L. Moore, Chief of Weather Bureau, Professor Cleveland Abbe, Editor. United States Department of Agriculture, Weather Bureau, Washington, D. C. 10 cents a copy.

An invaluable publication for teachers and students alike. Contains complete meteorological summaries for each month; accounts of all notable storms, cold and hot waves, etc.; and a large number of articles on a wide range of meteorological subjects. The charts show the tracks of areas of high and low pressure which crossed the United States during the month, the total precipitation, sea-level pressure, temperature and surface winds, percentage of sunshine, etc., for the month. Other charts are also frequently added.

The Journal of School Geography. Professor Richard E. Dodge, Teachers College, Columbia University, New York City, Editor. Publication Office, 41 No. Queen Street, Lancaster, Pa. Ten numbers a year. $1.00 per annum.

A monthly journal devoted to the interests of the common school teacher of geography. Contains numerous articles and notes on meteorological and climatological subjects.

Science. Edited by Professor J. McK. Cattell, Columbia University. New York City. New York, The Macmillan Company. Weekly. $5.00 per annum.

APPENDIX B.

Devoted to the advancement of all sciences. Contains brief *Current Notes on Meteorology*, which summarize the more important meteorological publications.

Monthly Bulletins of the Climate and Crop Service of the Weather Bureau.

These *Bulletins* are issued every month at the central office of the Weather Bureau in each State, under the direction of the Section Director of the Climate and Crop Service in that State. They contain meteorological data for the month, and frequently notes of interest. The annual summaries are especially valuable.

E. Charts.

Daily Weather Maps. These are published at the central office of the Weather Bureau in Washington, and at eighty-four other stations of the Bureau throughout the United States. It is best to have the daily maps sent from the nearest map-publishing station, and not from Washington, as the delay in the latter case is often so great that much of the immediate value of the maps is lost.

Climate and Crop Bulletin of the United States Weather Bureau. Washington, D. C. Monthly.

Chart showing, by means of small maps, the actual precipitation, departures from normal precipitation, departures from normal temperature, and maximum and minimum temperatures. Also a printed summary of the weather and of the crop conditions in the different sections of the United States. Issued on the first of each month.

Snow and Ice Chart of the United States Weather Bureau. Washington, D. C. Weekly during the winter season.

Based on data from regular Weather Bureau stations, supplemented by reports from selected voluntary observers. Shows, by shading, the area covered with snow at 8 P.M. each Tuesday during the winter, and by lines, the depth of snow in inches. Explanatory tables and text accompany the chart.

Storm Bulletin of the United States Weather Bureau. Washington, D. C. Issued at irregular intervals.

Charts, with text, illustrating the history of individual notable storms.

Pilot Chart of the North Atlantic and North Pacific Oceans. Hydrographic Office, Bureau of Equipment, Department of the Navy, Washington, D. C. Monthly. Price 10 cents a copy.

Shows calms and prevailing winds, ocean currents, regions of fog and equatorial rains, the positions of icebergs and wrecks, steamship and sailing routes, storm tracks, magnetic variation, etc. Also gives isobars and isotherms and a forecast for the month succeeding the date of publication, and a review of the weather over the oceans for the preceding month. Supplementary charts are occasionally issued.

Rainfall and Snow of the United States as compiled to the End of 1891, with Annual, Seasonal, Monthly, and other Charts. MARK W. HARRINGTON. United States Department of Agriculture, Weather Bureau, Bulletin C, Washington, D. C. 1894. Atlas, 18 x 24 inches. Charts 23. Text, 4-80 pp.

Contains twenty-three charts as follows: Monthly rainfall, seasonal rainfall, annual rainfall, monthly snowfall, monthly maxima of rainfall, rainy seasons, details of rainfall, details of occurrence of thunderstorms. Well adapted to serve as illustrations for use in the class-room. The text is explanatory, and is published separately in quarto form.

Rainfall of the United States, with Annual, Seasonal, and other Charts. ALFRED J. HENRY. United States Department of Agriculture, Weather Bureau, Bulletin D, Washington, D. C. 1897. 9¼ x 11½ inches. Pp. 58. Charts 10. Plates III.

A more recent publication than the preceding one, the averages having been compiled to the end of 1896. The charts are smaller than most of those in Bulletin C, and therefore not so well adapted for class-room illustration. The chart of mean annual precipitation is the latest and best published. The rainfall of the crop-growing season receives separate treatment, and is illustrated by means of two charts. The discussion in the text is excellent.

F. METEOROLOGICAL TABLES.

Smithsonian Meteorological Tables. Smithsonian Miscellaneous Collections, 844. Washington, D. C. 1893. 8vo. Pp. 262.

A very complete set of tables.

APPENDIX B. 193

Handbook of Meteorological Tables. H. A. HAZEN (of the United States Weather Bureau). Washington, D. C. 1888. 8vo. Pp. 127. $1.50.

Contains forty-seven tables, comprising all that are needed by the working meteorologist. Includes tables for Fahrenheit and Centigrade conversions, for barometric hypsometry and reduction to sea level, for the psychrometer, etc.

Tables for Obtaining the Temperature of the Dew-Point, Relative Humidity, etc. United States Department of Agriculture, Weather Bureau, Washington, D. C. 1897. 8vo. Pp. 29.

These are the tables now in use by the Weather Bureau.

G. ILLUSTRATIONS.

Classification of Clouds for the Weather Observers of the Hydrographic Office. Hydrographic Office, Bureau of Navigation, Department of the Navy, Washington, D. C. 1897. Sheet of twelve colored views. Price 40 cents. In book form, with descriptive text, $1.00.

An excellent set of cloud views, classified according to the *International Nomenclature*. The text describes the various cloud forms and shows their value as weather prognostics. An attractive addition to the furnishings of a school-room.

Selected List of Cloud Photographs and Lantern Slides.

Consists of twenty-eight photographs, and the same number of lantern slides, of the typical cloud forms, selected by the present writer from the collection in the Physical Geography Laboratory of Harvard University. The photographs (20 cents each, mounted) and slides (40 cents each) may be purchased of E. E. Howell, 612 17th Street, N. W., Washington, D. C. A description of these views was published in the *American Meteorological Journal* for July, 1894 (Boston, Mass., Ginn & Company).

Photographs. Photographs of miscellaneous meteorological phenomena, such as snow and ice storms, damage by storm-waves or high winds, wind-blown trees, lightning, etc., may often be purchased of local dealers. They add to the attractiveness of a school-room and furnish excellent illustrations in teaching.

H. GENERAL.

The following *Bulletins* of the Weather Bureau may be found useful as reference books:

No. 1. *Notes on the Climate and Meteorology of Death Valley, California.* MARK W. HARRINGTON. 8vo. 1892. Pp. 50.

No. 8. *Report on the Climatology of the Cotton Plant.* P. H. MELL. 8vo. 1893. Pp. 68.

No. 10. *The Climate of Chicago.* H. A. HAZEN. 8vo. 1893. Pp. 137.

No. 11. *Report of the International Meteorological Congress held at Chicago, Ill., Aug. 21-24, 1893.* 8vo. Pt. I, 1894, pp. 206. Pt. II, 1895, pp. 583. Pt. III, 1896, pp. 772. Pt. IV, not yet issued.

No. 15. *Protection from Lightning.* ALEXANDER McADIE. 8vo. 1895. Pp. 26.

No. 17. *The Work of the Weather Bureau in Connection with the Rivers of the United States.* WILLIS L. MOORE. 8vo. 1896. Pp. 106.

No. 19. *Report on the Relative Humidity of Southern New England and Other Localities.* A. J. HENRY. 8vo. 1896. Pp. 23.

No. 20. *Storms, Storm Tracks and Weather Forecasting.* FRANK H. BIGELOW. 8vo. 1897. Pp. 87.

No. 21. *Climate of Cuba. Also, A Note on the Weather of Manila.* W. F. R. PHILLIPS. 8vo. 1898. Pp. 23.

No. 23. *Frost: When to expect it and how to lessen the Injury therefrom.* W. H. HAMMON. 8vo. 1899. Pp. 37.

No. 25. *Weather Forecasting: Some Facts Historical, Practical, and Theoretical.* WILLIS L. MOORE. 8vo. 1899. Pp. 16.

No. 26. *Lightning and the Electricity of the Air.* In two parts. A. G. McADIE and A. J. HENRY. 8vo. 1899. Pp. 74.

The following miscellaneous publications of the Weather Bureau may also prove of value.

Injury from Frost and Methods of Protection. W. H. HAMMON. 8vo. 1896. Pp. 12.

Some Climatic Features of the Arid Regions. WILLIS L. MOORE. 8vo. 1896. Pp. 19.

APPENDIX B. 195

Investigation of the Cyclonic Circulation and the Translatory Movement of the West Indian Hurricanes. The late REV. BENITO VIÑES, S. J. 8vo. 1898. Pp. 34.

Requests for weather maps, *Bulletins*, and other publications of the Weather Bureau should be sent to the Chief of the Weather Bureau, Washington, D. C. All requests are dealt with on their merits, and in cases where it is deemed that effective use will be made of the publications they are usually sent free of charge.

INDEX.

A.

Anemometer, 38-41.
Aneroid barometer, 23, 24.
Anticyclones, 76.
—— and pressure changes, 141.
—— and temperature, 104-106.
—— and weather, 109-112.
—— and wind circulation, 98-100.
—— form and dimensions of, 96-98.
—— progression of, 111, 112.
—— tracks of, 111, 112.

B.

Backing winds, 118.
Barograph, 36, 37.
—— records, 37, 38.
Barometer, aneroid, 23, 24.
—— corrections, 33, 34.
—— mercurial, 19-23, 32, 33.
—— reduction to freezing, 143, 144, 166, 167.
Barometer reduction to sea level, 144, 145, 168-170.
Buran, 75, 76.
Buys-Ballot's Law, 99.

C.

Cherrapunjee, rainfall at, 138.
Clouds and upper air currents, 136, 137.
Clouds as weather prognostics, 137.
—— forms and changes of, 136.
—— movements of, 41-43, 46, 95, 136, 137.
Clouds, observations of, 5, 9, 41-43, 45, 46.
Cold wave, 62, 75, 76, 103, 106.

Cold-wave forecasts, 63.
Correlations of weather elements, 91-113.
Cyclones, 76.
—— and pressure changes, 141.
—— and temperature, 104-106.
—— and weather, 109-111, 113, 114.
—— and wind circulation, 98-100.
—— form and dimensions of, 96-98.
—— progression of, 111, 112.
—— tracks of, 111, 113.
—— tropical, 106.
—— velocity of, 112, 113.

D.

Dew, 134, 135.
—— point, 30.
—— —— tables, 142, 143, 146-155.
Diurnal variation of temperature, 125-127.
Diurnal variation of wind velocity, 130.
Doldrums, 29.

E.

Equipment of a meteorological laboratory, 186.
Evaporation, 29.

F.

Fahrenheit, 12, 13.
Ferrel's Law, 93.
Forecasts, cold-wave, 63.
—— weather, 49, 114-124.
Franklin, Benjamin, 114, 115.
Frost, 135.
—— warnings, 135.

INDEX.

G.

Galileo, 12, 19.
Gradients, pressure, 82-85.
—— temperature, 64-70.
—— vertical temperature, 129.

H.

Humidity, 29, 132-134.
—— diurnal variation of, 133.
—— relative, 30, 32, 45, 133.
—— relative, and wind direction, 133.
—— tables, 143, 156-165.
Hurricane, 100.

I.

Inversions of temperature, 129.
Isobaric charts, 76-82, 85, 95, 96.
Isobars, 77-82, 121.
Isothermal charts, 63, 68-70.
Isotherms, 55-57, 121.

L.

Land and sea breezes, 3, 131, 132.
Loomis, 97, 113.

M.

Meteorological tables, 146-170.
Mistral, 75.
Monsoons, 100.
Mountain and valley winds, 131.

N.

Nephoscope, 41-43.

O.

Observational meteorology, problems in, 125-141.
Observations, advanced instrumental, 26-40.
Observations, cloud, 5, 9, 41-43.
—— elementary instrumental, 11-26.
Observations, non-instrumental, 1-10.
Observations, rainfall, 9, 10, 26.
—— state of sky, 5, 9.
—— temperature, 1-3, 25.
—— wind, 3-5, 8, 9, 25.

P.

Pampero, 76, 104.
Pascal, 21.
Precipitation, 7, 9, 10, 138, 139.
Pressure and wind direction, 91-93.
—— and wind velocity, 93-96.
—— atmospheric, 19-21.
—— charts, 76-83.
—— cyclonic variation of, 140, 141.
—— decrease with altitude, 139, 140.
—— diurnal variation of, 140, 141.
—— gradient, 82-85.
Prevailing westerly winds, 93, 95, 96, 112, 113.
Psychrometer, 28.
—— sling, 31.
Purga, 75.

R.

Rain, see Precipitation.
—— heavy, 138, 139.
—— gauge, 15-17.
Rainfall records, 17, 18, 26.

S.

Sensible temperatures, 29, 30.
Sirocco, 103, 104, 133, 134.
Sling psychrometer, 31.
Smudges, 135.
State of sky, 5, 9, 45.
Suggestions to teachers, 171.

T.

Temperature charts, 58-60.
—— distribution, 51-63.
—— diurnal range, 125-127.
—— forecasts, 116.
—— gradient, 64-70.

Temperature gradient, vertical, 129.
—— inversions, 129.
—— maximum and minimum, 45.
—— mean, 43, 45.
—— observations, 1–3, 8–10, 25, 43, 45.
—— range, 45, 125–127.
—— sensible, 29, 30.
—— vertical distribution of, 128, 129.
Thermograph, 34–36.
—— records, 35, 36.
Thermometer, 12.
—— attached, 33.
—— maximum and minimum, 26–28.
—— shelter, 13, 14.
—— wet and dry bulb, 28, 30, 31.
Torricelli, 19, 20.
Trade winds, 93.

V.

Veering winds, 118.
Vernier, 33.
Vertical temperature gradient, 129.

W.

Water vapor, 28.
Weather, 85–90.
—— and wind direction, 106–109.
—— changes, sequence of, 113, 114.
—— charts, 85–90.

Weather forecasts, 49, 114–124.
—— map data, 90.
—— maps, 47–51.
—— prognostics, 137.
—— signs, 85.
—— temperate zone, 88–90.
—— torrid zone, 88–90.
Wind charts, 70–75.
—— direction and pressure, 91–93.
—— —— and relative humidity, 133.
—— —— and temperature, 101–104.
—— —— and weather, 106–109.
—— —— forecasts of, 118.
—— observations, 3–5, 25, 40, 41.
—— rose, 103, 108, 109.
—— vane, 15.
—— velocity, 3, 40, 41, 45.
—— —— and pressure, 93–96.
—— —— diurnal variation of, 130.
—— —— forecasts of, 118.
—— —— scale, 3.
Winds around cyclones and anticyclones, 98–100.
Winds, mountain and valley, 131, 132.
—— prevailing westerly, 93, 95–96, 112, 113.
Winds, trade, 93.
Woeikof, 69.